# THE WAR ON HUMANS

# Books by Wesley J. Smith

*The Lawyer Book: A Nuts and Bolts Guide to Client Survival*

*The Doctor Book: A Nuts and Bolts Guide to Patient Power*

*The Senior Citizens Handbook: A Nuts and Bolts Guide to More Comfortable Living*

*Winning the Insurance Game* (co-authored with Ralph Nader)

*The Frugal Shopper* (co-authored with Ralph Nader)

*Collision Course: The Truth About Airline Safety* (co-authored with Ralph Nader)

*No Contest: Corporate Lawyers and the Perversion of Justice in America* (co-authored with Ralph Nader)

*Forced Exit: Euthanasia, Assisted Suicide and the New Duty to Die*

*Culture of Death: The Assault on Medical Ethics in America*

*Consumer's Guide to a Brave New World*

*A Rat is a Pig is a Dog is a Boy: The Human Cost of the Animal Rights Movement*

# The War
## on Humans

WESLEY J. SMITH

DISCOVERY INSTITUTE PRESS   SEATTLE   2014

## Dedication
To Dean and Gerda Koontz, exceptional humans and cherished friends.

## Description
The environmental movement has helped produce significant improvements in the world around us—from cleaner air to the preservation of natural wonders such as Yellowstone. But in recent years, environmental activists have arisen who regard humans as Public Enemy #1. In this provocative book, Wesley J. Smith exposes efforts by radical activists to reduce the human population by up to 90% and to grant legal rights to animals, plants, and Mother Earth. Smith argues that the ultimate victims of this misanthropic crusade will be the poorest and most vulnerable among us, and he urges us to defend both human dignity and the natural environment before it is too late.

## About the Author
Named by *National Journal* as one of America's leading experts in the area of bioethics, attorney Wesley J. Smith is a Senior Fellow of Discovery Institute's Center on Human Exceptionalism and the previous author of books such as *A Rat is a Pig is a Dog is a Boy: The Human Cost of the Animal Rights Movement*, *Consumer's Guide to a Brave New World*, and *Culture of Death: The Assault on Medical Ethics in America*. Smith also writes the popular *Human Exceptionalism* blog at *National Review*.

## Library Cataloging Data
*The War on Humans*
Wesley J. Smith
110 pages, 6 x 9 x 0.27 inches & 0.4 lb, 229 x 152 x 0.67 cm. & 0.18 kg
Library of Congress Control Number: 2014955945
BISAC: POL044000 Political Science/Public Policy/Environmental Policy
BISAC: NAT039000 Nature/Animal Rights
BISAC: NAT011000 Nature/Environmental Conservation & Protection
BISAC: POL035010 POLITICAL SCIENCE/Human Rights
BISAC: SCI075000 Science/Philosophy & Social Aspects
ISBN-13: 978-1-936599-26-4 (Paperback)
ISBN-13: 978-1-936599-17-2 (Kindle)
ISBN-13: 978-1-936599-16-5 (EPub)

## Publisher Information
Discovery Institute Press, 208 Columbia Street, Seattle, WA 98104
Internet: http://www.discoveryinstitutepress.com/
Published in the United States of America on acid-free paper.
First Edition, First Printing. November 2014.
Cover Design: Brian Gage. Interior Layout: Michael W. Perry

# ACKNOWLEDGMENTS

The seed of this book was planted by Discovery Institute Founder and Chairman Bruce Chapman. When I was first made a Senior Fellow while researching *A Rat is a Pig is a Dog is a Boy: The Human Cost of the Animal Rights Movement*, Bruce handed me a file he had been keeping and suggested that once I completed the animal rights book, I might want to take a look at how human exceptionalism was under attack from radical environmentalists in the Deep Ecology movement. Bruce was not only correct in his concern about Deep Ecology, but in the nearly ten years since that conversation, the once radical enviromental fringe has chewed its way steadily into the mainstream—as I hope this book convincingly demonstrates.

I am deeply grateful to Bruce and my other colleagues at the Discovery Institute who have unfailingly supported my work and shared its purposes. Particular thanks to Steve Buri for his unflagging backing and John West for his publishing and editing magic. A tip of the hat to the technical team who formatted the book in the mysterious ways of on-line publishing and to Cameron Wybrow, the copy editor. Thanks also to Steve Meyer, David Klinghoffer, Richard Sternberg, Ann K. Gauger, Rob Crowther, Janine Dixon, Anna Selick, Casey Luskin, Jonathan Wells, Jay Richards, and Jens Jorgenson.

My sympathies to my friends who put up with my obsessions: Tom Lorentzen, Bruce and Valerie Scholley, John Miller and Stephanie Brown, Dan and Jennifer Lahl, William and Erica Hurlbut, Mark and LaRee Pickup, Fr. Leo and Denise Arrowsmith, Bradford Short, Rita and Mike Marker, John and Kathi Hamlon, and Joseph and Lorena Bottum.

As ever, my love to the Saunders family, South Carolina, Florida, New York, Connecticut, and Rhode Island branches: Jerry and Barbara, Jim and Vickie, Jennifer, Jeremiah and Sara and their sons, Patrick, Connor, and Aidan, Stephen and Leslie, Rebecca and Jonathan Shulman, Eric, and Joshua. My mother, Leona Smith, who continues to amaze, and most of all, to the source of so much joy, my wife and total sweetheart, Debra J. Saunders.

# Contents

# Introduction: People Are the Enemy

In 1972, Canadian science broadcaster David Suzuki told some giggling students, "One of the things I've gotten off on lately is that basically... we're all fruit flies." But that was just the start: Suzuki then likened us to "maggots" who are "born as an egg" and "eventually hatch out and start crawling around," eating and "defecating all over the environment."[1]

Denigrating humans as maggots was edgy back in the hippy-dippy days (and Suzuki looked the part with his long-hair and John Lennon-style glasses), but few took such assertions very seriously. They were made to shock or get attention more than to express genuine misanthropy. Back then, the environmental movement didn't generally denigrate human beings. Rather, it advocated preventing and cleaning up pollution, protecting endangered species, and conservation as a matter of human duty. Those are noble goals, ones which I support.

Unfortunately, the primary values of the original environmental movement have gone the way of bell-bottom jeans. In recent years, like termites boring into a building's foundation (to borrow a Suzuki-type metaphor), anti-humanism has degraded environmentalist thinking and advocacy. Indeed, environmental activists today routinely denigrate humans as parasites, viruses, cancers, bacteria, and murderers of the Earth.

Suzuki, now a world-famous celebrity and anti-global warming activist, certainly hasn't changed his old anti-human views. When asked by a Canadian Broadcasting Corporation interviewer in 2009 about how his "not very optimistic" perception of humanity has changed since he called people maggots, Suzuki merely deflected the question, noting that racism had lessened but also lamented that "Humanity is humanity... I just wish they'd stop being so human!"[2]

The popular culture has certainly embraced Suzuki's anti-humanist theme. The A-List remake of *The Day the Earth Stood Still*, starring the movie mega-star Keanu Reeves, provides a vivid case in point. In the 1951 version, "Klaatu," a space alien played by a mild Michael Rennie, lands in a flying saucer on a goodwill mission to save humanity from itself.[3] The space community, he tells world leaders, is aware of humanity's growing technological prowess and wants to welcome us warmly into the interstellar club; however, worried about our warlike ways, they have assigned Klaatu to warn us that we must become peaceable—or else. If we threaten to spread violence into space, we risk destruction from the killer robot "Gort."

Like all good science fiction, *The Day the Earth Stood Still* used a fantastical premise to warn about contemporary cultural concerns. The film was made just as the Cold War was growing white hot. Korea had exploded into war. The Soviet Union and the USA were developing ever more powerful nuclear weapons. People were terrified that humanity might wipe itself out in a nuclear apocalypse. The movie reflected the overwhelming view that *humanity was not only worth saving,* but also, that we could rescue ourselves if only we followed the better angels of our nature.

That was then. The message of the remake is not only far, far darker, but seethes with antipathy toward the human race. The remake's alien community not only has no interest in saving humanity; it doesn't think we are worth the effort. To the contrary: Klaatu has mutated from a benign friend of man into a dangerous and brooding enemy. Gort the killer robot isn't a potential threat to our future, but a weapon to be immediately deployed against us for total annihilation. Why? Not because the space civilization worries we will bring war to the stars, but because total genocide is the only way to *save the Earth.*

Klaatu justifies making humans extinct in very blunt terms, with the clear implication that the Earth itself is a living entity:

**Klaatu:** This planet is dying. The human race is killing it.

**Helen Benson:** So you've come here to help us.

**Klaatu:** No, *I* didn't.

**Helen Benson:** You said you came to save us.

**Klaatu:** I said I came to save the Earth.

It is also worth noting that Klaatu's flying saucer has been replaced by a blue sphere that looks like a mini-planet complete with clouds. Other Earth-looking spheres soon arrive, and it quickly becomes clear that they are extraterrestrial Noah's arks, tasked with removing all other species from Earth to be returned to thrive unmolested after the great human obliteration.

As in the original, Klaatu is shot, captured, interrogated, and then escapes. He is befriended by a boy and his mother, who are unaware (at first) that their new friend is the fugitive space alien. They and a scientist who won a Nobel Prize for "biological altruism"—played straight-faced by Monty Python's John Cleese—manage to convince Klaatu that humanity is worth preserving. After all, we have classical music! But before he can call off the genocide, Gort transforms into a huge nano-swarm and swiftly spreads, obliterating all in its path.

At the last possible moment, Klaatu cancels our extinction, but he warns us that our salvation will come at a significant cost: As his space sphere leaves the atmosphere, it emits a pulse disabling all technology on Earth. The implications are clear: In order to co-exist peacefully with the planet, we must return to a non-technological state. Unmentioned in this "happy ending" is the hard truth that such a sudden technological collapse would kill billions from starvation and disease.

Unfortunately, *The Day the Earth Stood Still* isn't the only recent A-list Hollywood movie to preach the toxic idea that humans deserve to be wiped out for the supposedly unconscionable harm we are doing to the biosphere. *The Happening*, starring Mark Wahlberg, offers an apocalyptic tale of a rebellion against the oppressive human hegemony—*by plants!*[4]

Written and directed by noted filmmaker M. Night Shyamalan, *The Happening* opens with people on the East Coast suddenly committing mass suicide. The images are startling as civilization quickly collapses. We eventually discover that plants have mounted a great floral rebellion in which they release "suicide pheromones" whenever humans gather in large groups, forcing all to immediately kill themselves by any means handy—lying in front of a mechanical lawn mower, crashing cars into walls, jumping off buildings, whatever.

Shyamalan, as he did in the alien invasion film *Signs*,[5] tells his apocalyptic story from the micro perspective of Wahlberg's family and friends as they cope with the chaos. As mass suicide spreads throughout the Northeast, Wahlberg's small band is steadily pushed into ever tighter corners. The small group takes refuge in a model home in a new housing development. Realizing that plants release the suicide pheromones whenever a critical mass of human beings is present—and that the number of people necessary to stimulate the release is shrinking—Wahlberg and company flee at the approach of a larger group of refugees. As members of the larger group begin to kill themselves *en masse*, Wahlberg *et al.* run past a huge advertising sign for the housing development that carries the unsubtle message of the film: "Because you deserve it."

The attacks finally end. But lest anyone think it was a fluke, an ecological expert advises on television that the plants have sent us a warning to change our ways—or else! Cut to Paris, as people begin to kill themselves on the *Champs Elysées*.

It is some comfort that neither of these misanthropic films did well at the box office. However, the fact remains that the ever-avaricious Hollywood studio system believes there is an anti-human entertainment market to be mined, spending hundreds of millions of dollars and countless hours of some of the industry's most creative talent to preach the darkest dogmas of radical environmentalism around the world. (The mega-hit *Avatar* is another example.[6]) Indeed, the appearance of A-list Hollywood anti-human message films demonstrates that environmental nihilism has moved beyond the fringe and has begun—like the pheromones in *The Happening* —to infect the cultural mainstream.

Misanthropy as an advocacy strategy would have once raised eyebrows. Alas, as we shall see, it is now par for the environmental course.

These anti-human and anti-technology attitudes are not restricted to fiction. Nor are they restricted to psychopaths like the notorious Ted Kaczynski, of Unabomber infamy.[7]

Take the seemingly benign "Earth Hour," which mounts an annual worldwide campaign urging people to turn off lights and electric appliances for 60 minutes. I had always thought of Earth Hour in the same way I do those ubiquitous colored ribbon-in-the-lapel advocacy campaigns—essentially, it allows people to feel good about themselves for caring, without actually becoming self-sacrificing.

But then I read the following column by Ross McKitrick in the *Vancouver Sun* and realized that the message—if not the explicit intent—of Earth Hour is profoundly destructive:

> The whole mentality around Earth Hour demonizes electricity... Earth Hour celebrates ignorance, poverty and backwardness. By repudiating the greatest engine of liberation it becomes an hour devoted to anti-humanism. It encourages the sanctimonious gesture of turning off trivial appliances for a trivial amount of time, in deference to some ill-defined abstraction called "the Earth," all the while hypocritically retaining the real benefits of continuous, reliable electricity.
>
> People who see virtue in doing without electricity should shut off their fridge, stove, microwave, computer, water heater, lights, TV and all other appliances for a month, not an hour. And pop down to the cardiac unit at the hospital and shut the power off there too.[8]

Some might say, "But Wesley, Earth Hour is only about publicizing the need to be efficient with energy use." Perhaps, but I think McKitrick is onto something.

Consider the broader context of an environmentalism that is becoming increasingly nihilistic, anti-modern, and anti-human, to the point that even the noble concept of conservationism has become passé because it implies the propriety of exploiting resources to create wealth and prosperity. As we will explore in the following pages, the anti-human side of today's environmental movement has many manifestations:

- The deep ecology movement that would decimate human population to under one billion.
- The global warming alarmists' tyrannical and anti-growth tendencies.
- The ludicrous notion of granting legal and enforceable legal "rights" to nature.
- The concomitant international campaign to criminalize large-scale resource exploitation and land development projects as "ecocide."

Now look again at the core message of "Earth Hour." That message is not subtle: Technology is the enemy of the Earth and must be severely curtailed in order to "save the planet"—just as Klaatu curtailed technology to save the Earth in the movie.

One wonders how many proponents of Earth Hour have seriously considered the human misery that exists where there are no power grids, where life is precarious and often suffused with profound suffering from hunger and disease. In reality, for the love of suffering humanity, we *need more electricity, not less*. Perhaps those considering turning off their light switches next year should first ask the suffering people of North Korea how they like living in a substantially non-technological society.[9]

Anti-humanism is only part of the problem we face from the green misanthropes. This book will also explore the extent to which radical environmentalism is corrupting scientific rationalism—perhaps with malice aforethought. Indeed, it is very clear to me that ideology seeks to supplant the dispassionate scientific method of obtaining and applying facts and data—come what may—with an emotional fervor reminiscent of a quasi-religious movement seeking to impose its dogma into policy and law.

People are noticing the corruption, undermining the population's general trust in the scientific sector. In 2011, for example, *Scientific American* ran a hand-wringing article seeking to explain why "people say that they trust scientists in general but part company with them on specific issues?"[10]

Allow me to help. The public is smart enough to differentiate between what are sometimes called bench scientists, and the politicized "science" advocates, the latter of whom too often strive to conflate our overwhelming support for science per se as agreement with their favored political/ideological agendas. Moreover, people have come to understand that scientific studies are like Scripture: You can read into them almost anything you want. Too often the conclusions the scientists want seems to precede their actual study of the evidence.

And then there is the attempt by some self-appointed "science advocates" to corrupt and co-opt the scientific method as a justification for a misguided philosophy known as scientism. Scientism mistakenly asserts that science can not only tell us the way things are and how things work, but also identify right from wrong. Too often in the environmental area, scientists have moved from revealing the objective to promoting the subjective. As this book will make clear, that is when matters often go seriously off the rail.

We will also explore how the Green Movement has become brown. What do I mean? "Progressive" (Red) agendas too often subsume legitimate environmentalist (Green) concerns. When red is mixed with green, it creates brown. By allowing itself to become explicitly anti-capitalist and free market in its sensibilities and messaging, the Green Movement has browned, which—along with ideological zeal—explains the growing inclination of environmentalists to embrace authoritarian solutions.

To cite an example: A well-known environmental author named David Shearman wrote a hysterical piece in 2008 urging that we embrace worldwide authoritarianism to force implementation of the "scientific consensus" on global warming and other so-called international environmental emergencies. His shining example of progressive environmental governance? *The People's Republic of China*—you know, the country of forced abortion, Tiananmen Square, organ selling from executed prisoners, etc. No matter. Shearman asserted that China should be extolled because it *banned plastic bags!* According to Shearman:

> Liberal democracy is sweet and addictive and indeed in the most extreme case, the USA, unbridled individual liberty overwhelms many of the collective needs of the citizens. The subject is almost sacrosanct and those

who indulge in criticism are labeled as Marxists, socialists, fundamentalists and worse. These labels are used because alternatives to democracy cannot be perceived!...

The Chinese decision on shopping bags is authoritarian and contrasts with the voluntary non-effective solutions put forward in most Western democracies. We are going to have to look how authoritarian decisions based on consensus science can be implemented to contain greenhouse emissions... If we do not act urgently we may find we have chosen total liberty rather than life.[11]

Ah, the old "Let us rule the world or we are all going to die" gambit.

Even our old pal David Suzuki—who thinks people are maggots—recently joined the Green Authoritarian Chorus, opining that politicians be *jailed* for violating the scientific consensus on global warming. Suzuki told a McGill University audience: "What I would challenge you to do is to put a lot of effort into trying to see whether there's a legal way of throwing our so-called leaders into jail because what they're doing is a criminal act. It's an intergenerational crime in the face of all the knowledge and science from over 20 years."[12]

Never mind that Marxist governments have been history's worst ravagers of the environment. And never mind that democratic societies cleaned up their polluting ways because that is what the people wanted. And never mind that authoritarianism never achieves permanently benign ends.

Freedom is not a luxury. Human flourishing requires liberty and prosperity in which to thrive, the latter good being dependent on land use and resource development. Such activities are now under pronounced threat—a development that can be traced back to the Deep Ecology Movement of the 1970s. And that is where we will begin.

# 1. Deep Anti-Humanity

DURING ITS FIRST CENTURY—FROM EVEN BEFORE THE HALCYON days of John Muir and Theodore Roosevelt—environmentalism succeeded brilliantly. In 1872, United States President Ulysses S. Grant signed a law establishing Yellowstone as the world's first national park. Today, there are 1,200 national parks and sanctuaries around the world, and that doesn't include state and local preservation efforts.

There is no doubt that the drive for economic growth, for a time, led to poor environmental practices. But it didn't last. Reacting against the filthiness of the mid-twentieth century, people and governments from around the world confronted unacceptable industrial pollution that fouled air and water, and they began to reform industrial standards to reduce emissions and set about the important task of cleaning up past messes—a process that continues to this day.

These efforts worked. I grew up in a Los Angeles choking on smog so thick you couldn't see the nearby San Gabriel Mountains. Yet, within a few years we were breathing easier and again able to see the local topography.

Elsewhere cleanup efforts stopped rivers from catching fire, made lakes habitable for fish and other wildlife again, and remediated toxic waste dumps. In short, the environmental movement contributed mightily to the making of a better, healthier, and cleaner world.

But beginning in the late 1960s, a subversive misanthropy began to gestate within environmentalism, a view that did not see the Earth and the fullness thereof—in the Biblical turn of phrase—as ours to develop for human benefit. Rather, it castigated people as a disease infecting the planet, best treated with the antibiotic of massive human depopulation and opposition to development and economic growth.

## Deep Ecology Gives Birth to the Anti-Human Movement

The Deep Ecology Movement blazed the trail to environmentalist anti-humanism. The term was coined by Norwegian philosopher Arne Dekke Eide Næss, who, inspired by the environmental alarmism of Rachael Carson's *Silent Spring*, rejected human exceptionalism, arguing that each facet of the natural world—including humanity—is equal to every other. Moreover, he insisted, each element is entitled to receive the same consideration in the way we live our individual lifestyles and in the public policies we pursue. In other words, humans are duty-bound to care as much for the rest of the natural world as we do for ourselves—even if doing so comes at significant human expense.

Unlike the ideas that characterized mainstream environmental thinking up until that era—denigrated by Næss as "shallow ecology" because it generally accepted the propriety of responsibly developing resources and making beneficial use of land—Næss demanded that we *forgo material thriving* in order to make common cause with flora and fauna. He also embraced radical Malthusianism,[13] insisting that our world population be *reduced to 100 million*—today, there about 7 billion of us. This is an impossible number to attain absent genocide, a catastrophic die-off caused by a pandemic, or the total collapse of technology.

Over the years, Næss and his co-believers developed Deep Ecology into a social movement with a list of specific ideological goals published in 1984 with George Sessions. At the time, these goals were quite radical, but as we shall see, they have become mainstream within contemporary environmentalism:

### *The Deep Ecology Platform*

1. The well-being and flourishing of human and nonhuman life on Earth have value in themselves. These values are independent of the usefulness of the nonhuman world for human purposes.

2. Richness and diversity of life forms contribute to the realization of these values and are also values in themselves.

3. Humans have no right to reduce this richness and diversity except to satisfy vital needs.

4. Present human interference with the nonhuman world is excessive, and the situation is rapidly worsening.

5. The flourishing of human life and cultures is compatible with a substantial decrease of the human population. The flourishing of nonhuman life requires such a decrease.

6. Policies must therefore be changed. The changes in policies affect basic economic, technological, and ideological structures. The resulting state of affairs will be deeply different from the present.

7. The ideological change is mainly that of appreciating life quality (dwelling in situations of inherent worth) rather than adhering to an increasingly higher standard of living. There will be a profound awareness of the difference between big and great.

8. Those who subscribe to the foregoing points have an obligation directly or indirectly to participate in the attempt to implement the necessary changes.[14]

Read the platform carefully. If we are allowed to exploit resources only to fulfill "vital needs," our existence would again be lived largely at the mercy of nature—which for most of our history, was an all-too-grim reality in which most human lives were desperate, brutal, and short. Had our forefathers and foremothers acted as if the natural world had essentially equal value "independent" of its "usefulness... for human purposes," we would have largely thwarted such modern advances as industrialization, electricity, mechanized travel, and the like. Indeed, our drive to improve human existence created the material prosperity, vastly improved health, and far longer lives enjoyed by the developed world.

Our continuing task should be to make these tremendous advances universally available to our still-struggling brothers and sisters in the developing world. But implementing the values of Deep Ecology would thwart that noble goal, amounting to an abrupt about-face, disassembling most of what we have heretofore accomplished. In this sense, rather than a being a progressive step toward a more enlightened existence as it pretended, Deep Ecology was actually profoundly reactionary, rejecting modernity outright, willing to return us to the bad old days of the raw struggle for survival.

But there is more going on than a loathing of industrial development and concomitant urbanization. Deep Ecology would create a neo-Earth religion as the grounding of a new human moral system—steeped in anti-capitalism and rejecting technology. Thus, the Deep Ecology Foundation's mission statement contends that the "current problems"—e.g. the supposed despoiling of the natural world—"are largely rooted in the following circumstances":

+ The loss of traditional knowledge, values, and ethics of behavior that celebrate the intrinsic value and *sacredness of the natural world* and that give the preservation of Nature prime importance…

+ The prevailing economic and development paradigms of the modern world, which *place primary importance on the values of the market*, not on Nature…

+ *Technology worship and an unlimited faith in the virtues of science;* the modern paradigm that technological development is inevitable, invariably good, and to be equated with progress and human destiny.[15] [Emphasis added.]

In short, Deep Ecology is profoundly misanthropic. It not only devalues human worth, but because it rejects core human advances—science, technology, economic growth—that liberated us from so much want and deprivation.

From all reports, Næss was a gentle soul who sincerely believed that his backward-looking views would improve the human condition. The same cannot be said of many contemporary green anti-humanists in whose hands the core attitudes of Deep Ecology have become aggressively antagonistic to humanity.

## Gaia Theory

At about the same time Næss was conjuring Deep Ecology, another environmentalist philosopher named James Lovelock posited an equally radical idea, which has become known as Gaia Theory: The Earth (pagan goddess Gaia) "evolved... as a single living, self-regulating system."[16] Almost as if the planet is intelligent, Gaia Theory argues that "life maintains conditions suitable for its own survival," and it urges us to treat the environment—more accurately, the Earth—as a living being: "the living system of Earth can be thought of analogous to the workings of any individual organism that regulates body temperature, blood salinity, etc."

A 2006 column published in the *San Francisco Chronicle* by University of Arizona professor Albert J. Bergesen vividly exposed the Deep Ecology humans-are-equal-to-nature ideology undergirding Gaia Theory. Bergesen argued that we should think of ourselves as "eco-beings." Rejecting human exceptionalism outright, he essentially declared humanity the moral equal of rocks, spider, fungi, trees, plankton, squirrels—and indeed, all of nature.

Bergesen's ecoegalitarianism also advocated quasi-Earth religion mysticism:

> To continue to focus upon realizing our humanity only mystifies our true eco-being, for we were part of nature before we took on the manifestations of being human. The categorical location of consciousness as human, or animal, and perhaps as even plant or rock, river, or mountain, may be merely an accident of Gaian birth.

> Think of this: If we believe we are ultimately human beneath the social categories we occupy, why aren't we basically eco-beings beneath the species categories we occupy? And if the earlier humanistic goal was to realize our full species being, why can't the new environmental goal be to realize our eco-being? *As long as we continue to identify only with our humanity, we will continue in a state of eco-alienation, and the more we push exclusively for human concerns, the more we deny our true eco-nature.* [Emphasis added.][17]

What would that mean in practical terms? Bergesen doesn't say, so allow me: If fungi and ants are equal to people, we are ethically required to subjugate human welfare to ensure they receive equitable treatment in recognition of their supposedly equal moral worth. This idea—expressed in hard neo-Darwinian terms by some, and in neo-mystical terms by others (as by Bergesen)—goes well beyond the belief in human/animal moral equality as posited by the animal rights movement. To paraphrase PETA co-founder Ingrid Newkirk's most infamous statement: "A rat, is a pig, is a dog, is a boy, is a beetle, is the Potomac River." (See my *A Rat is a Pig is a Dog is a Boy: The Human Cost of the Animal Rights Movement* for a detailed exposition and criticism of animal rights ideology.[18])

Gaia Theory meshes seamlessly with Deep Ecology. Both reject human exceptionalism. Both diminish the intrinsic importance of humanity. Both redefine us as merely one equal member (to borrow a Christian metaphor)—with no greater or lesser moral significance—of Gaia's body. Both see environmentalism as a much higher and nobler cause than promoting human flourishing. And both, with the sheer gravitational force of human logic, lead to profoundly misanthropic values, principles, and consequences.

Given the similarity in their approaches and philosophical destinations, it is unsurprising that Lovelock was an early believer in global warming and embraced Deep Ecology's potentially catastrophic radical human depopulation meme as the remedy. Thus, on page 154 of his 2006 book, *The Revenge of Gaia*, Lovelock sighs that "we should listen to the deep ecologists and let them be our guide":

> This small band of deep ecologists seem to realize more than other green thinkers the magnitude of the change of mind needed to bring us back to peace within Gaia, the living Earth. *Like the holy men and women who make their whole lives a testament to their faith,* the deep ecologists try to live as a Gaian example for us to follow. [Emphasis added.][19]

If actually implemented, Gaia Theory and Deep Ecology would lead to the collapse of modern civilization. Cutting billions from the human population—the genocidal implications of which are seemingly of little concern—makes sense *if one sees Gaia as an organism and humans as a dangerous pathogen.* Thus a book reviewer noted the tyrannical turn in Lovelock's advocacy:

"If we are to continue as a civilization that successfully avoids natural catastrophes, we have to make our own constraints on growth and make them strong and make them now." As it is, we are unintentionally at war with Gaia and must agree to "wartime" rationing and temporary "loss of freedom."

*Strong constraints? Loss of freedom?* What's the difference between this and ecofascism? And how far down will our population have to plummet to satisfy Gaia? Actually, Lovelock states that something like nine-tenths of our population must vanish: "Personally I think we would be wise to aim at a stabilized population of about half to one billion." To accomplish this goal, both the birth rate and death rate would have to be "regulated" as "part of population control." So we are to be bred, managed, and put down just like a herd of animals on a farm. If this isn't totalitarianism, what is?[20]

Deep Ecology and Gaia Theory—once radical outliers that could be dismissed merely as the kook fringe—have become respectable. Alas, as we shall see, the tyrannical impulses of the early theorists have grown like a gathering storm. And the deluge is almost upon us.

# 2. Homo Sapiens, Get Lost

WHEN THE KIND OF NIHILISM THAT ANIMATES DEEP ECOLOGY/ Gaia Theory are on the fringe, it doesn't matter much. But when anti-humanism overflows the dikes of human exceptionalism, it could bring a flood of woe, threatening Western concepts of liberty and our material prosperity.

Human exceptionalism and its corollaries—such as universal human rights—have been integral to Western progress. Over the last few hundred years, the moral foundations of society became progressively pro-human. Judeo-Christian moral philosophy and secular humanism both promoted human flourishing and the protection of individual rights as primary purposes of society.

But in recent decades green misanthropy has chewed its way steadily from the edges into the soft center of the environmental movement. Influenced by the anti-humanistic and capitalism-hostile values of Deep Ecology, environmentalism began to project its advocacy beyond conserving resources, preserving uniquely pristine areas, promoting proper industrial practices, and protecting endangered species, into a movement that either expressly or implicitly seeks to reduce human prosperity and freedom in the name of "saving the planet."

It's only logical: As flora and fauna came to be perceived as equivalent in moral value to humans, it didn't take long to brand human exceptionalism "arrogant" and harmful to nature. Identifying ourselves as the villains, in turn, opened the door to wide acceptance of the demoralizing nihilism that makes a virtue out of equating humanity to a vermin infestation or a virus infecting the planet.

For example, Paul Watson, the head of the Sea Shepherd Conservation Society—famous for its angry demonstrations against Canadian seal pup harvests and aggressive anti-whaling campaigns—has repeatedly expressed deep antipathy toward humans. Demonstrating that the Society's motives and purposes extend far beyond saving the lives of cute seal pups and leviathan, in 2007 Watson unleashed a strident anti-human rant on his Web site:

> We are the ruthlessly territorial primates whose numbers have soared far beyond the level of global carrying capacity for the deadly behavioral characteristics that we display. This did not happen yesterday because we suddenly became aware of the dangers of global warming. It began 50,000 years ago when a relatively hairless primate stumbled out of equatorial Africa and began wiping out the megafauna of the time.[21]

Watson depicts humans as the worst form of disease:

> I was once severely criticized for describing human beings as being the "AIDS of the Earth." I make no apologies for that statement. Our viral like behavior can be terminal both to the present biosphere and ourselves. We are both the pathogen and the vector. But we also have the capability of being the anti-virus if only we can recognize the symptoms and address the disease with effective measures of control.

Watson warns that the treatment required to cure this disease will be harsh: "Curing a body of cancer requires radical and invasive therapy, and therefore, curing the biosphere of the human virus will also require a radical and invasive approach."

Actively pursuing policies consistent with such human-loathing would not "save the planet." But it certainly could provide moral justification to the unhinged to act out against the human virus in a violent manner. Indeed, we have already seen murderous terrorism motivated by the supposed harm human technology has done "to the natural world" (Unabomber Ted Kaczynski).[22] Similarly, radical Malthusian James Lee—killed during a violent hostage- taking—wanted to "keep the human race from breeding any more disgusting babies."[23]

Watson is hardly alone among activists who hope for—or, more charitably, who *expect*—profound ills to fall upon the human race as a matter of the Earth's self-defense. Most notoriously, University of Texas evolutionary biology professor Eric R. Pianka was accused of yearning for an Ebola or similar viral pandemic with a 90 percent kill rate to cure human overpopulation. Pianka denied the charge, claiming he actually *predicted* that potential catastrophe, rather than hoping for it to come to pass. Since the speech was only partially recorded, we will probably never know whether he hoped for human near-extermination or merely warned that it was coming.

Yet, even in the published version of Pianka's speech, he clearly advocates the deconstruction of modern civilization (my emphasis):

> Our economic system based on continual growth must be replaced by a sustainable system where each of us leaves the planet in the same condition that it was in before we were born. *This will require many fewer of us and much less extravagant lifestyles. We won't be able to move around so freely (airplanes will become a thing of the past) and we will have to go back to walking and riding horses.* In addition, humans will have to be more spread out, living without big cities. Before it is all over, we are going to have to limit our own reproduction, un-invent money, control human greed, revert back to trade and barter, and grow our own crops, among other things.[24]

Again, shades of *The Day the Earth Stood Still*: Pianka's anti-technology advocacy would cause billions of human casualties.

Regardless of whether he actually *wants* humans wiped out by Ebola, Pianka cannot escape the charge of being an anti-human. Accusing "creationists from the lunatic fringe" of slandering him (including a few of my Discovery Institute colleagues, who are not creationists), the professor took to the Internet to set the record straight. It was an odd defense to the charge of misanthropy that reveals an abiding anti-human bigotry that so typifies contemporary Deep Ecology ideologists:

> Most of my talk was an appeal for respect of endangered species and natural habitats, and I deplored the all too common anthropocentric attitude among many humans that we are above nature and that we can do whatever we want with natural habitats and other species.[25]

To the contrary—human exceptionalism actually expresses a directly contrary view. Rather than licensing despoliation, it imposes a *positive duty* on us to properly manage the environment precisely *because* we are human. But let's not interrupt Pianka's self-incriminating rant:

I compared the brainless runaway population growth of humans to bacteria growing exponentially on an agar plate. To underscore my point, I said that "we are no better than bacteria!"

Here's the actual quote: "Humans are no better than bacteria, in fact, we are just like them when it comes to using up resources... We are *Homo* the *sap*, not *sapiens* (stupid, not smart)." That's even more insulting than David Suzuki's old charge that we are "maggots."

Oh well, some might assure themselves, that's just fringe talk. *Au contraire!* Perhaps illustrating the extent of politicization within contemporary science, Pianka gave his speech after being named the Distinguished Texas Scientist of 2006.

## Draconian Depopulation

Green anti-humanism has also unleashed a draconian Malthusianism that seeks to radically reduce the number of human beings in the world. Before we go any further, please be clear that I am not criticizing voluntary family planning, which I recognize helps many to live better, more prosperous lives. While reasonable people may differ profoundly over family planning *methods*—issues beyond our scope here—determining the number and timing of children offers substantial potential for improving family welfare, as well as bringing desperately needed relief from profound want in the world's poorest areas.

But providing people with the means to decide the number of children to have isn't the same thing as limiting family size through the force of law or radical depopulation schemes—my focus here. As we shall see, advocates in very high places advocate profoundly immoral population polices in the name of Green.

Jonathan Porritt, the former chair of the UK's Commission on Sustainable Development, promoted increased abortion as a tool to save the planet—even to the point of sacrificing the well-being of the sick to pay for the policy's costs. The *Times of London* reported:

Couples who have more than two children are being "irresponsible" by creating an unbearable burden on the environment, the government's green adviser has warned. Jonathon Porritt, who chairs the government's Sustainable Development Commission, says curbing population growth through contraception and abortion must be at the heart of policies to fight global warming... Porritt, a former chairman of the Green party, says the government must improve family planning, even if it means shifting money from curing illness to increasing contraception and abortion.[26]

The rise in anti-human values also has fueled a concomitant flirtation with anti-freedom. For example, in the question-and-answer portion of his notorious speech, Eric Pianka praised the tyrannical government of the People's Republic of China for its "one child" population control efforts: "The reason China was able to turn the corner and is gonna become the new super power in the world is because they got a police state and they can force people to stop reproducing."[27] If one believes we are an infection on the Earth, tyranny provides a logical means for containing the pathogen.

Alas, as we shall see throughout this book, Pianka is not a lone voice, despite China's policy having been repeatedly exposed as involving forced abortions, rampant female infanticide, and the unleashing of an anti-disability eugenics. As described in the *Wall Street Journal*, author Susan Greenhalgh detailed the consequences in her book, *Cultivating Global Citizens: Population in the Rise of China*:[28]

> In the quest for "superior" children and mothers, Ms. Greenhalgh explains, Beijing put aside social, cultural and political factors, and discriminated against whole classes of low-quality people: "rural residents, rural migrants to the cities, women, minorities, and those with substandard bodies" as well as "deviants" such as homosexuals and unmarried couples. She writes, "Some 'low-quality' citizens (such as women) have been targeted for energetic, state-sponsored improvement campaigns, while others (rural people for example) have been essentially abandoned as useless to the modernization effort."[29]

Eugenics is also official Chinese policy, and the population has fallen into blatant sex selection:

> Official pressure to curtail and "improve" births resulted in infanticide and selective abortion, which in turn led to a gender gap among newborns of at best 120 boys to 100 girls. Many Chinese men, therefore, will not find brides, and fewer elderly Chinese will have daughters to comfort and

support them. Many urban Chinese have internalized the bias against the family and, as described by Ms. Greenhalgh, feel little obligation to care for their parents, want no children and think only of getting rich.

It is stunning how little such cultural corruption matters to those caught up in the frenzy of Green.

The meme that human beings are the enemy of the planet finds its most extreme voice in the Voluntary Human Extinction Movement.[30] Arguing that "Our collective consciousness must evolve from anthrocentric to eco-centric: to where Earth has first priority," the VHEM Website sighs ecstatically about a world without humans:

> Sounds like paradise, doesn't it? Gaia completely cured of pox humanus. Without us meddlesome humans, all other species would get their fair chance at survival.

> Naturally, it's not that simple, but just for fun, let's envision an impossible dream: all human sperm suddenly and permanently loses viability—no impregnated human egg begins meiosis to form a zygote—none transforms from embryo into the sacred fetus, is carried to term and sentenced to life. Zero conceptions wanted nor un[dertaken]… Benefits would begin immediately for both biosphere and humanity. Resources wasted on redundant breeding could be redirected to existing members of the human family in need. Loving care and nurturing now expended raising superfluous heirs could be given over to stopping the killing and beginning the healing. A sweet dream.[31]

Don't laugh. Books have been written about the paradise that would emerge after we were extinguished. Of course, no one would know, because the remaining species would be incapable of appreciating a sunset or the sheer joy of walking through a pristine forest.

## THE DARK MOUNTAIN PROJECT: EMBRACING "UNCIVILIZATION"

THE DARK MOUNTAIN PROJECT IS A GOOD EXAMPLE OF WHAT TOMORROW's environmentalism may look like. Formed by Paul Kingsnorth,[32] former *Ecologist* deputy editor, and social activist Paul Hine, the Dark Mountain Project seeks to harness the power of art to embrace economic decline and break the back of society's belief in human exceptionalism.

In a 2011 Q & A interview in the *Ecologist*—which describes Dark Mountain as "the most influential strand of green thought to emerge within the last five years"—Kingsnorth explained what the DMP hopes to accomplish. It ain't pretty:

> Dark Mountain is a process. First you give up on the unrealistic ambitions of both the mainstream growth narrative and the mainstream green narrative—creating a "sustainable" consumer democracy for nine billion people, stopping climate change, making capitalism nice and all the rest of the utopian stuff. You give up on that and then you can move on. You embrace what we might call radical honesty, and you look at what is possible and what's not, and you see your true place in the world and in history... We call this the "hope beyond hope." First you get real. Then you give up. Then you are re-inspired.[33]

Maybe it's my age, but that's just gobbledygook to me. So, I went to the Dark Mountain Website to see if I could find a better explanation. There, I found the "Eight Principles of Uncivilization," among which is:

> 3. We believe that the roots of these crises lie in the stories we have been telling ourselves. We intend to challenge the stories which underpin our civilization: *the myth of progress, the myth of human centrality*, and the myth of our separation from 'nature'. These myths are more dangerous for the fact that we have forgotten they are myths. [Emphasis added.][34]

These so-called "myths" have led over the last several hundred years to more freedom—from tyranny, from want, and from suffering— than any time in human history. But never mind, there is a planet to save:

> 5. Humans are not the point and purpose of the planet. Our art will begin with the attempt to step outside the human bubble. By careful attention, we will reengage with the non-human world.[35]

Ah, the nonsense that we are merely another animal in the forest. We'll be seeing more of that as we go along.

Elsewhere, Kingsnorth comes across as the Green equivalent of the stereotypical fundamentalist Christian yearning for the Apocalypse. In an exchange with environmentalist George Monbiot, published in the *Guardian*, he writes:

Some people... believe that these things should not be said, even if true, because saying them will deprive people of "hope", and without hope there will be no chance of "saving the planet". But false hope is worse than no hope at all. As for saving the planet—*what we are really trying to save, as we scrabble around planting turbines on mountains and shouting at ministers, is not the planet but our attachment to the western material culture,* which we cannot imagine living without. The challenge is not how to shore up a crumbling empire with wave machines and global summits, but to start thinking about how we are going to live through its fall, and what we can learn from its collapse. [emphasis added][36]

How disappointed do you think Kingsnorth will be if the downfall doesn't come?

Perhaps as disturbing was Kingsnorth's interlocutor, the supposedly more reasonable Monbiot. Monbiot agrees that collapse is inevitable, but he fights on to prevent total human catastrophe:

The human and ecological consequences of the first global collapse are likely to persist for many generations, perhaps for our species' remaining time on earth. To imagine that good could come of the involuntary failure of industrial civilization is also to succumb to denial. The answer to your question—what will we learn from this collapse?—is nothing.

This is why, despite everything, I fight on. I am not fighting to sustain economic growth. I am fighting to prevent both initial collapse and the repeated catastrophe that follows. However faint the hopes of engineering a soft landing—*an ordered and structured downsizing of the global economy*—might be, we must keep this possibility alive. [Emphasis added.][37]

In other words, planned, pronounced, and severe economic downturn is our last best chance. Good grief!

It is tempting to dismiss such existential despair. But there's a real danger here: If enough people reject the very notion of progress, indeed, if we embrace "uncivilization," the very collapse Kingsnorth claims is inevitable and which Monbiot hopes to mitigate could become a self-fulfilling prophecy.

## The Georgia Guide Stones Urge Worldwide Deep Ecology

A while back, I gave a speech highlighting the threats to human exceptionalism. I brought up my concern about Deep Ecology's call to reduce humankind to under 1 billion. During the Q and A, an audience member asked me what I thought of the Georgia Guide Stones. I had never heard of them, but promised to investigate.[38]

I learned that the Guide Stones were huge message-bearing granite monoliths erected by unknown persons in 1979. Written in different languages, part of the Stones' advocacy is merely feel-good pabulum, such as promoting one world government and occultism, issues with which we need not concern ourselves here.

But some of the Stones' directives appear to support eugenics and the imposition of a legal policy of Deep Ecology that would require mass human extermination—issues that go to the heart of my concerns about the current trends in environmentalism:

1. Maintain humanity under 500,000,000 in perpetual balance with nature.

2. Guide reproduction wisely—improving fitness and diversity…

10. Be not a cancer on the earth—Leave room for nature—Leave room for nature.

Alarmingly, the Guide Stones have found friends in influential places. As just one example, in 2013 a *Discover* magazine contributing editor Jill Neimark extolled the monument for its supposed visionary message:

> As I stood there before them I thought of our present realities of climate change, soaring global population, and the devastation of species. We were 3.4 billion when the Guide Stones were erected. We are now 7 billion headed to 10 billion by 2050. "Be not a cancer on this earth," the tenth and final inscription on the stones commands. "Leave room for nature. Leave room for nature." That afternoon, those rough-hewn giants could not have been more eloquent.[39]

Eloquent? Sure, if you hate humanity.

The warm embrace of the Guide Stones in a *science magazine* is a symptom of a deep anti-human fever that infects contemporary environmentalism. Indeed, in the next chapters, I will show that these ideologies—like a plague virus in a Michael Crichton novel—have escaped their test tubes, infecting the heart of the mainstream.

# 3. Global Warming Hysteria

We now turn our attention to present-day environmental misanthropic madness. Unavoidably, that brings us to the most penetrating and pressing contemporary environmental controversy: global warming—or as it is now often called, climate change.

I am not a climate scientist. Indeed, I have no formal scientific training, and thus I am not about to express a categorical opinion on whether the Earth is warming, and if so, the extent to which that is caused by human activities. Moreover, this book is not intended to take particular policy positions. Rather, I hope to expose how anti-humanism and self-loathing have corrupted the noble goal of promoting environmentally responsible practices—to the point that we may sacrifice our own thriving on the altar of "saving the planet!"

Still, in the interests of serving that over-used and under-utilized concept, "transparency," I believe it is only right and proper to disclose my personal opinion on the larger question. Essentially, I am agnostic. (How's that for guts?) Unlike global warming's most vocal critics, I don't think that all of those worried about climate change are perpetrating an elaborate "hoax." Indeed, the evidence shows that the Earth did warm in the last century, and I think, there are reasons to believe that humans could have had some impact in that regard. Moreover, some very smart people whom I respect a lot believe that we are, indeed, causing climate change, and there is no question that some of the researchers in the skeptics' camp are for hire by fossil fuel industries. I know how that polka is danced.

This isn't to say the alarmists are pure. Indeed, there is far more money to be had in promoting global warming alarmism than fighting it. Al Gore certainly cashed in from his global warming alarmism. And accepting the threat of climate change is necessary to obtain grant funding at the university and government consulting levels: There's gold in *them-thar* global warming hills!

The anti-global warming cause has also been substantially co-opted by anti-capitalism ideologues, and it has been used as an excuse to stifle heterodox findings and theories. The meme that "the scientific consensus" should settle the matter completely not only contradicts the scientific method, but corrupts the way science is supposed to be conducted.

So, to the point: I have paid close attention to the arguments. There is no dispute that carbon dioxide levels have increased in the last one hundred years. There has been some warming in recent decades.

But does that equal causation? Perhaps not. Despite the steady and ongoing increases in carbon, there hasn't been a statistically significant increase in world temperatures in nearly twenty years.[40] Nor do we know precisely what part solar cycles play in this whole thing, not to mention cloud cover, ocean currents, etc. Climate is driven by a system of almost unimaginable complexity, currently beyond our complete understanding.

I also have noted that most of the early computer-generated climate models predicting ongoing warming have not proved valid. Moreover, we have seen dramatic temperature variations in Earth's history—indeed in recorded times—that had nothing to do with human beings. Nor am I convinced that we face such a dire climate crisis or that it is reasonable to say we are "killing the planet." I suppose that puts me mildly in the skeptics' camp.

But for my purposes here, it doesn't matter. I am not seeking to convince anyone about the truth or folly of either side. Nor will I parse the different studies offered by the climate scientists. Rather, in the next chapter, I hope to underscore the extent to which subversive anti-humanism has percolated into the premier environmental issue of our times.

## GLOBAL WARMING HYSTERIA

BEFORE WE GET TO THE EXPLICIT ANTI-HUMANISM, I THINK IT WOULD be worthwhile to take a brief detour into the extreme hyperbole unleashed by advocates to convince a reluctant world of the righteousness of the warming cause.

Why is this important? Sustaining robust anti-humanism requires a perennial crisis. Inspiring hatred of self requires a parade of horribles. We certainly have that when it comes to global warming alarmists. They offer up a never-ending parade of supposedly imminent catastrophes that will destroy us unless we immediately cap the greenhouse gas spigot. In other words, to keep from killing the planet, we have to cease and desist from most of the normal human activities and enterprises that have made modern life so rich and fulfilling.

It's become a joke, really, and I don't want to take up too much of your time recounting all of the silly—and often contradictory—hysterical claims made by global warming propagandists over the years. But they're just too juicy to let go:

*The Himalayan Glaciers Will Disappear by 2035!* Wrong! In 2007, the United Nations Intergovernmental Panel on Climate Change (IPCC) warned darkly that the Himalayan Glaciers would disappear *by 2035.* Shocking! Think about the farm lands below that will be flooded! Oops, their mistake. In 2009, the IPCC retracted the claim.[41]

*The Polar Bears Are Going to Drown!* Wrong! In an age in which emotional narratives often trump facts, the polar bear became the icon of global warming hysteria. After a report about four drowned polar bears seen far out at sea, they were put on the USA threatened list—even though there has been no demonstrated diminution in population. Al Gore put the polar bears in his propaganda movie, *An Inconvenient Truth.* Children are still warned that if they don't cut their carbon footprints, the polar bears will drown.

The alarmists are going to have to come up with a new icon of catastrophe. The original study was discredited and indeed, recent investigations demonstrate that the polar bear population is not only thriving, but actually has grown in the last forty years.[42]

*Global Warming Is Melting the Snows of Kilimanjaro!* Wrong! Al Gore and other hysterics claimed that the "snows of Kilimanjaro," made famous by Hemingway, would soon disappear as global warming melted the glaciers on the extinct volcano. But it turns out deforestation was the actual cause of reduced snow cover.[43] Oh, and in recent years, the snows have been coming back.[44] Oh well, next case.

*Global Warming Will Stop the Snow!* Wrong! The IPCC put out a report in 2001 predicting that "milder winters will decrease heavy snowstorms." Some advocates even claimed that children in England would grow up not knowing what snow was. Since then, there have been several winters with heavy snow.[45] So now the hysterics are warning that global warming may bury us in snow! Heads we win, tails you lose.

*We Only Have [Fill in the Number] Years to Save the Planet!* Scream in abject fear! Run around in circles! Store food! In 2011, we were warned that we only have *ten years* to save the planet!: "The world has just ten years to bring greenhouse gas emissions under control before the damage they cause become irreversible, the Met Office has warned."[46] But we heard the *very same warning* back in *2005*:

> Global warming is reaching the point-of-no-return, with widespread drought, crop failure and water shortages the likely result, according to a new international report... The countdown to climate-change catastrophe is spelt out by a task force of senior politicians, business leaders and academics. In 10 years or less, they predict, the catastrophic point-of-no-return may be reached.[47]

Oh dear! And then there was a story back in 2009 reporting that NASA's head hysteric, James Hansen, believed we only have four years left— e.g., until 2014![48] *We only have 4 years!!!*

The ever-hysterical Al Gore agreed with the ten years left hypothesis— *back in 2006.*[49] And a newspaper report from 2007 warned that we only have until 2015.[50] These guys should get their stories straight! The moral of the story? In global warming hysteria, there is always a looming crisis requiring us to give up our freedoms.

*Emitting Greenhouse Gases Is Like Bombing Bangladesh!* A radical in one field generally thinks radically in other areas of ideological engagement. Witness utilitarian bioethicist Peter-let's-permit-infanticide-Singer, who also has weighed in on global warming. Predictably, Singer's comments drip with hysteria:

> Suppose we were waging aggressive war on Bangladesh, let's say, and we were dropping lots of bombs and somebody said we should stop this war, 'we have no justification for declaring war on Bangladesh'… And somebody else said 'well, but Australia has a big industry manufacturing bombs and if we stop the war we'll harm the economy because there won't be all these jobs'. Now I don't think even the [evil] conservatives would support that argument but what we're doing is not really very different. Now that we know the effects of our greenhouse gas emissions, we are harming people in Bangladesh almost as surely as if we were dropping bombs on them.[51]

Bangladesh is not facing "inundation" any time soon even if there is global warming. But it is currently mired in terrible poverty. The real harm would be to disable our economies, which would have a cascading effect, including increasing suffering to countries like Bangladesh, both by making less aid available and by inhibiting such countries from developing their own resources.

*Our Grandchildren Will Not Live Into Old Age!* Australian microbiologist, Dr. Frank Fenner, whose work helped defeat smallpox, made world headlines in 2010 when he predicted that we are all doomed because of global warming, telling *The Australian*: "Homo sapiens will become extinct, perhaps within 100 years. A lot of other animals will, too. It's an irreversible situation."[52] Needless to say, some icons of popular culture—who can be counted on to swallow the hysteria whole as good career moves—push the agenda. For example, in 2013, singer Sir Bob Geldof warned that all humans would be dead by 2030 because of the potential impact of climate change.[53]

*There Will Be an Ice-Free Arctic by 2014!* Al Gore predicted direly that "new" information coming to his attention suggests an ice-free arctic by summer 2014. Oh no! Stifle all economic activity! Return to being hunter/gatherers—better yet, just gatherers! One little problem: The professor, upon whose work Gore's hysteria was based, says he didn't actually predict that. From the story:

Speaking at the [2009] Copenhagen climate change summit, Mr Gore said new computer modeling suggests there is a 75 per cent chance of the entire polar ice cap melting during the summertime by 2014. However, he faced embarrassment last night after Dr Wieslav Maslowski, the climatologist whose work the prediction was based on, refuted his claims. Dr Maslowski, of the Naval Postgraduate School in Monterey, California, told *The Times:* "It's unclear to me how this figure was arrived at. I would never try to estimate likelihood at anything as exact as this."[54]

Worse, for the global warmers, in 2013, arctic ice had increased dramatically.[55] What's a hysteric to do?

***Global Warming Is Killing the Lizards!*** Scientists have issued a study warning that global warming will cause mass lizard extinction. From the *Reuters* story:

Scientists warn in a research paper published on Thursday that if the planet continues to heat up at current rates, 20 percent of all lizard species could go extinct by 2080. "The numbers are actually pretty scary," said lead researcher Barry Sinervo from the University of California Santa Cruz. "We've got to try to limit climate change impacts right now or we are sending a whole bunch of species into oblivion." A mass extinction of lizards, which eat insects and are eaten by birds, could have devastating effects up and down the food chain, but the extent is difficult to predict.[56]

Wait a minute: The "current rate" of warming is negligible—even Professor Phil Jones, the prominent University of East Anglia climate change champion—acknowledged that there has been no "statistically significant" warming for the last 15 years.[57]

***Kill Your Dog to Save the Planet!*** *They hate animals too!* Two global warming hysterics have argued that owning a dog is a threat to the planet. From a column in the *New Scientist:*

Should owning a great Dane make you as much of an eco-outcast as an SUV driver? Yes it should, say Robert and Brenda Vale, two architects who specialise in sustainable living at Victoria University of Wellington in New Zealand. In their new book, *Time to Eat the Dog: The Real Guide to Sustainable Living,* they compare the ecological footprints of a menagerie of popular pets with those of various other lifestyle choices—and the critters do not fare well...

Over the course of a year, Fido wolfs down about 164 kilograms of meat and 95 kilograms of cereals. It takes 43.3 square metres of land to generate 1 kilogram of chicken per year—far more for beef and lamb—and 13.4 square metres to generate a kilogram of cereals. So that gives him a footprint of 0.84 hectares...Meanwhile, an SUV—the Vales used a 4.6-litre Toyota Land Cruiser in their comparison—driven a modest 10,000 kilometres a year, uses 55.1 gigajoules, which includes the energy required both to fuel and to build it. One hectare of land can produce approximately 135 gigajoules of energy per year, so the Land Cruiser's eco-footprint is about 0.41 hectares—less than half that of a medium-sized dog.[58]

Oh no! *Dogs are killing the planet!* What is a good radical environmentalist to do?

"Shared pets are the best – the theatre cat or the temple dogs," says Robert Vale. But if you must own your own, think about getting an animal that serves a dual purpose. He recommends hens, which partly compensate for their eco-footprint by providing eggs. Or there is an even better alternative, if you can stomach it. "Rabbits are good," he says, "provided you eat them."

The joy dogs give us can't be measured in gigajoules. Oh well, if this craziness is to be stopped, getting between people and their beloved pets is the sure way to do it.

***Hurricanes Are Proof of Global Warning!*** Hurricane Irene—not a huge storm as these things go—caused extensive damage in 2011 when it took an unusual, but certainly not unprecedented, path up the East Coast's most populated areas. That set off some typical hysteria. From GWH Central, aka, the *New York Times:*

The scale of Hurricane Irene, which could cause more extensive damage along the Eastern Seaboard than any storm in decades, is reviving an old question: are hurricanes getting worse because of human-induced climate change? The short answer from scientists is that they are still trying to figure it out. But many of them do believe that hurricanes will get more intense as the planet warms, and they see large hurricanes like Irene as a harbinger.[59]

Harbinger? The gist of the story is that there is disagreement among the climate modelers, who don't have a good record of accuracy in any event.

And no one can say Irene was "caused" by global warming. But why let a good storm go to waste?

Please. The reason I am here to write these words—or better stated, the event that set the wheels in motion leading to my eventual birth—was the big northeastern hurricane of 1938. That storm, weighing in at a Category 3, was much stronger than Irene's Category 1. My mother and her family lived in Rhode Island. The storm so upset them that they traveled to balmy California to see if the state would appeal to them. They liked what they saw so much they picked up stakes and moved. That's when my mother met my father. And the rest, as they say, eventually became my history.

The USA was hit by zero hurricanes in 2013.[60] It will experience hurricanes again, of course. But the failure of the hurricane meme would seem to demonstrate that the storm warnings are over-blown.

***Global warming will cause earthquakes!*** The media have widely touted a calamitous future of increased earthquakes and volcanic eruptions due to global warming. Check out this story from the Associated Press reprinted on the CBS website:

> New research compiled by Australian scientist Dr. Tom Chalko shows that global seismic activity on Earth is now five times more energetic than it was just 20 years ago.

> The research proves that destructive ability of earthquakes on Earth increases alarmingly fast and that this trend is set to continue, unless the problem of "global warming" is comprehensively and urgently addressed.[61]

Many other news outlets also ran the story. Which raises an important question: Just who is Dr. Tom Chalko? Let's just say he's a, well, *exotic* character, who is into auras[62] and space aliens.[63] Chalko also believes that global warming could cause Earth to explode.[64]

Now space aliens and auras might be real, but someone who promotes their existence and worries that thermal imbalance will cause Earth to explode is hardly an "expert" whose opinions should be quoted by the AP and reported on CBS!

All the hysteria described above has generated surreal, not to mention hysterical, "curatives." For example, a bioethics professor named Matthew Liao proposed that we should *shrink*—literally—the human race:

> How could height reduction be achieved? Height is determined partly by genetic factors and partly through diet and stressors. One possibility is to use preimplantation genetic diagnosis, which is now employed in fertility clinics as a means of screening out embryos with inherited genetic diseases. One might be able to use preimplantation genetic diagnosis to select shorter children. This would not involve modifying or altering the genetic material of embryos in any way. It would simply involve rethinking the criteria for selecting which embryos to implant.

> Also, one might consider hormone treatment either to affect growth hormone levels or to trigger the closing of the growth plate earlier than normal. Hormone treatments are already used for growth reduction in excessively tall children.[65]

I doubt the good professor's proposal would find much support in the National Basketball Association.

Meanwhile, some Dutch scientists suggest that we begin eating bugs for our protein. As described in *Science's* website:

> Forget eating local or eating organic, the new way to dine green may be eating gross. Meat consumption presents a big environmental problem: Cows and pigs are good sources of protein, but they're also big sources of the heat-trapping gases carbon dioxide and methane, which they belch and, um, emit in other ways. One way to reduce such emissions while maintaining a nutritious diet may be to get people to eat more cricket burgers and mealworm patties. According to a new study, many insect species gain weight faster and spew out less greenhouse gases as they grow than do their beefier counterparts. The tricky part is getting them to look good on a menu.[66]

I am well aware, of course, that humans eat insects. St. John the Baptist ate locusts and honey as an ascetic discipline. But let's get real: People are not going to give up their sirloin steaks for leg of cricket. Nor, it seems to me, should we spend public resources (or private, for that matter, although that would be the funders' own business) studying whether to create a new industry in *Gryllidae* ranching.

We have been told that to save the planet from boiling we must not have children, dramatically reduce our prosperity, give up our cats and dogs, sweat in the summer without air conditioning, freeze in the winter with low heat settings, become vegan, eat bugs, stop traveling, etc., etc., etc. But the Japanese government suggested a relatively easy way out—just go to bed earlier. From the *Telegraph* story:

> The Japanese government has launched a campaign encouraging people to go to bed and get up extra early in order to reduce household carbon dioxide emissions. The Morning Challenge campaign, unveiled by the Environment Ministry, is based on the premise that swapping late night electricity for an extra hour of morning sunlight could significantly cut the nation's carbon footprint.[67]

Benjamin Franklin famously said, "Early to bed, early to rise, makes a man healthy, wealthy, and wise," even calling such a lifestyle a "religious duty." But he never dreamed his bromide would become a prescription for planetary salvation. Wait a minute: Hysterics tell us that to save the planet, the "wealthy" part will have to go. And greater poverty reduces health. It seems to me that if we give into global warming hysteria, that will prove we are not "wise." But, adopting that lifestyle could still be considered a religious duty as part of neo-Earth worship, so at least the entire quote might not have to be thrown in the dumpster.

Lest we think humans have only recently become planet-killers, one study blames global warming on our ancestral cave dwellers:

> Mammoths used to roam modern-day Russia and North America, but are now extinct—and there's evidence that around 15,000 years ago, early hunters had a hand in wiping them out. A new study, accepted for publication in *Geophysical Research Letters*, a journal of the American Geophysical Union (AGU), argues that this die-off had the side effect of heating up the planet. "A lot of people still think that people are unable to affect the climate even now, even when there are more than 6 billion people," says the lead author of the study, Chris Doughty of the Carnegie Institution for Science in Stanford, California. The new results, however, "show that even when we had populations orders of magnitude smaller than we do now, we still had a big impact."[68]

How did we do that? By killing the mammoths, we let birch trees take over grasslands, because the mammoths weren't there anymore to kill the saplings. (What about the "rights" of the trees against the pachyderms? Doesn't anybody care about that? Yes, they do, as a later chapter will demonstrate.)

The report on the study continues: "The trees would change the color of the landscape, making it much darker so it would absorb more of the Sun's heat, in turn heating up the air. This process would have added to natural climate change, making it harder for mammoths to cope, and helping the birch spread further." And that means we may have impacted the planet much earlier than thought:

> Earlier research indicated that prehistoric farmers changed the climate by slashing and burning forests starting about 8,000 years ago, and when they introduced rice paddy farming about 5,000 years ago. This would suggest that the start of the so-called "Anthropocene"—a term used by some scientists to refer to the geological age when mankind began shaping the entire planet—should be dated to several thousand years ago. However, Field and colleagues argue, the evidence of an even earlier man-made global climate impact suggests the Anthropocene could have started much earlier. Their results, they write, "suggest the human influence on climate began even earlier than previously believed, and that the onset of the Anthropocene should be extended back many thousands of years."

What are we supposed to do, clone mammoths and set herds free to restore the environment of eons ago?

If our ancient ancestors are guilty as charged, good for them! The warming temperatures made it possible for early humans to grow more food, populate greater portions of terrain, and eventually thrive to the point that we could step beyond naked evolution and exert substantial control over our environment—to the betterment of humankind.

All of this would be amusing, if it were not for the dark side of such advocacy. Perhaps foremost, all the hysteria is panicking our children. So in the Environmental Defense Fund's video advertisement "Tick," we see children warning darkly against "massive heat waves, severe droughts, devastating hurricanes," as they repeatedly say (or yell), "Tick!" The ad ends,

"Our future is up to you," requiring fighting of global warming "while there is still time."[69]

Scaring children about their futures is having a deleterious impact on the young. According to one study, one in three children now fear the world will end before they become adults. From the 2009 *Reuters* story:

> Has all of the attention on saving the planet these days actually created more anxiety about the state of the Earth for our children? Perhaps. A new survey finds that one out of three children ages 6–11 years old fear that the planet won't exist when they grow up and more than half (56%) believe that the Earth will not be as good a place to live. Minority children worried the most—with 75% of Black children and 65% of Hispanic children fearing the planet was going to deteriorate before they grew up.[70]

It turns out that the survey of 500 pre-teens was conducted by a radical environmentalist group called "Habitat Heroes,"[71] which inadvertently put a little truth in advertising in its press release: "Habitat Heroes is *targeted* towards children ages 6 to 14." Don't environmentalists know that it is cruel to terrify children?

Mental health professionals have also warned that fears over global warming are exacerbating the anguish experienced by some mentally ill patients:

> A recent study has found that global warming has impacted the nature of symptoms experienced by obsessive compulsive disorder patients. Climate change related obsessions and/or compulsions were identified in 28% of patients presenting with obsessive compulsive disorder. Their obsessions included leaving taps on and wasting water, leaving lights on and wasting electricity, pets dying of thirst, leaving the stove on and wasting gas as well as obsessions that global warming had contributed to house floors cracking, pipes leaking, roof problems and white ants eating the house. Compulsions in response to these obsessions included the checking of taps, light switches, pet water bowls and house structures. "Media coverage about the possible catastrophic consequences to our planet concerning global warming is extensive and potentially anxiety provoking. We found that many obsessive compulsive disorder patients were concerned about reducing their global footprint," said study author Dr Mairwen Jones.[72]

Panic-mongering, as we have seen, has thankfully not convinced the world to destroy economies and keep developing nations mired in low emissions destitution. But it has disturbed children and worried the mentally ill. Nice going.

# 4. Pardon Us For Living!

Reasonable people can differ on the persuasiveness of the evidence for man-caused global warming and the extent of danger that a warming of the planet might present. But there should be no disagreement that children should not be taught to hate humanity (in general), or to loathe themselves (in particular). But that is precisely the anti-human message often communicated by global warming hysterics in television ads and on Internet sites.

Take, for example, the video ad "No Pressure," created in support of a program called the "10:10 Campaign,"[73] which seeks to convince people to cut their individual carbon footprint by ten percent. In the ad, an elementary school teacher describes the program for her class and discusses possible planet-saving actions they could take—such as convincing their parents to take a train instead of a plane while traveling or using energy-saving light bulbs.

As the bell rings, she asks how many in the class are willing to do their part. All but two children raise their hands. She asks the class to wait a moment before leaving as she pushes a big red button: BLAM! The two dissenters explode so violently and graphically that their classmates and teacher are splattered with blood and broken flesh.[74]

Lest you think the vivid depiction of children being blown up as punishment for not fighting global warming was meant as satire, *The Guardian* reported:

> Our friends at the 10:10 climate change campaign have given us the scoop on this highly explosive short film, written by Britain's top comedy screenwriter Richard Curtis, ahead of its general release. It's most definitely striking and if you haven't watched it yet—taking into account the warning that it contains scenes some people may find disturbing—do so now, before I give too much away...

But why take such a risk of upsetting or alienating people, I ask her: "Because we have got about four years to stabilise global emissions and we are not anywhere near doing that. All our lives are at threat and if that's not worth jumping up and down about, I don't know what is. We 'killed' five people to make No Pressure—a mere blip compared to the 300,000 real people who now die each year from climate change," she adds.[75]

Global warming hysteria, apparently, justifies all! Needless to say, if conservative issue campaigners ever graphically depicted children being murdered as a punishment for their incorrect thinking, the screaming of media critics would never end.

The 10:10 Campaign has hardly been alone in symbiotically depicting the death of children as part of an answer to saving the planet. Until recently, the Australian Broadcasting Corporation's website carried a children's feature known as "Planet Slayer." The most offensive feature on the site was "Professor Schpinkee's Greenhouse Calculator"[76] a now-erased online game that told children how to *"find out when you should die!"*

This "game" asked children a series of questions about their lifestyles and consumption habits. Their answers were then computed. At the end, players were told the age they should die, having exhausted their individual share of the world's resources.

Once again, the visual involves explosions—not as vivid as the exploding flesh-and-blood children killed for not supporting the 10:10 Campaign—but gorily depicting a cartoon pig blowing up in a bloody mess. While an exploding cartoon pig may not be as offensive as a blown-up child, the imagery is in some ways more pernicious, because it communicates to children that *they are pigs* for consuming resources. Indeed, the game teaches children that they are planet-killers who should die before becoming adults. In fact, the grading was so tough—no mercy shown—that when I played the game Professor Schpinkee told me I should have died at age 7.4!

Remember, Planet Slayer was published on Australia's equivalent of the British Broadcasting Corporation's website—illustrating how mainstream anti-human global warming hysteria has become and how deeply it has penetrated Establishment institutions. As Australian columnist Andrew Bolt noted in the *Herald Sun*:

What a lovely insight into the green philosophy. Children should die to save the planet… A little joke, you will say. A mere attention grabber in a good cause. Trouble is, though, that there really is an insanely anti-human bent to deep green preaching on global warming, and there really are believers who feel only too keenly the planet is doomed by our sin, and humans must vanish.[77]

It's not nice to terrify children and make them believe that merely by living, they kill the planet.

Of course, some alarmists don't want to scare children into a lower-carbon lifestyle so much as cut the number of children in the world to terrify. As we have already seen, an influential adviser on environmental issues advocated in the *Times of London* that the UK adopt a two-child policy and increase the abortion rate as a method to reduce carbon dioxide output.[78]

Along the same lines, a study from the University of Oregon also supported the idea of being childless as a way to cut our carbon footprints. From the *Live Science* story:

> For people who are looking for ways to reduce their "carbon footprint," here's one radical idea that could have a big long-term impact, some scientists say: Have fewer kids. A study by statisticians at Oregon State University concluded that in the United States, the carbon legacy and greenhouse gas impact of an extra child is almost 20 times more important than some of the other environment-friendly practices people might employ during their entire lives—things like driving a high mileage car, recycling, or using energy-efficient appliances and light bulbs…
>
> Reproductive choices haven't gained as much attention in the consideration of human impact to the Earth, [Paul] Murtaugh said. When an individual produces a child—and that child potentially produces more descendants in the future—the effect on the environment can be many times the impact produced by a person during their lifetime.[79]

The good professors state they don't want any laws *requiring* small families. But the implication of their advocacy leads to that end, particularly since it is unlikely that enough people will voluntarily reduce the size of their families to make a substantial difference in carbon levels.

Illustrating just how whacky global warming Malthusianism can become, the Mother Nature Network published an article lauding Genghis Khan—the killer of millions of people—for wonderfully cooling the planet during his years of conquest. From the Mother Nature Network story:

> Genghis Khan's Mongol invasion in the 13th and 14th centuries was so vast that it may have been the first instance in history of a single culture causing man-made climate change, according to new research out of the Carnegie Institution's Department of Global Ecology, reports Mongabay. com. Unlike modern day climate change, however, the Mongol invasion cooled the planet, effectively scrubbing around 700 million tons of carbon from the atmosphere. So how did Genghis Khan, one of history's cruelest conquerors, earn such a glowing environmental report card? The reality may be a bit difficult for today's environmentalists to stomach, but Khan did it the same way he built his empire — with a high body count.
>
> Over the course of the century and a half run of the Mongol Empire, about 22 percent of the world's total land area had been conquered and an estimated 40 million people were slaughtered by the horse-driven, bow-wielding hordes. Depopulation over such a large swathe of land meant that countless numbers of cultivated fields eventually returned to forests. In other words, one effect of Genghis Khan's unrelenting invasion was widespread reforestation, and the re-growth of those forests meant that more carbon could be absorbed from the atmosphere.[80]

Only when the new Earth religion reigns can a vicious barbarian like Khan can be canonized a saint.

It's tempting to just laugh and roll our eyes. But here's the disturbing thing: Misanthropic environmentalism is moving ever deeper into the mainstream, and in fact, has been accepted whole cloth by many among what my generation used to call "The Establishment." Indeed, the weather has become an ever-ready excuse for imposing ever more intrusive and controlling policies intended to impede human flourishing, or tax human enterprise.

## INCREASE POVERTY TO COOL THE PLANET

GLOBAL WARMING HYSTERICS TALK ABOUT "GREEN JOBS" AND MAIN-taining human prosperity. But beneath it all, they embrace lowered standards of living and forcing the world's poorest areas into permanent destitution in order to "save the planet."

*Agenda: Poverty is the Answer to Global Warming.*

The willingness to sacrifice human welfare is reaching a fever pitch among those who believe that global warming is a crisis of unimagined proportions. An article by David Owen in the *New Yorker*—pushing the importance of economic decline to saving the planet—illustrates the point:

> The environmental benefits of economic decline, though real, are fragile, because they are vulnerable to intervention by governments, which, understandably, want to put people back to work and get them buying non-necessities again—through programs intended to revive ordinary consumer spending (which has a big carbon footprint), and through public-investment projects to build new roads and airports (ditto).[81]

The answer, apparently, is more of the same economic malaise we are now experiencing: "The ultimate success or failure of Obama's [anti-global warming] program, and of the measures that will be introduced in Copenhagen this year, will depend on our willingness, once the global economy is no longer teetering, to accept policies that will seem to be nudging us back toward the abyss."

In other words, people need to be poorer, with all the concomitant increase in human suffering and shorter lives that would cause. And remember, Owen only writes here about the well-off areas of the world. But you can bet that he and his co-believers would strive mightily to stifle development in now-destitute areas of the world—dooming perhaps billions of people to permanent dependency on the developed world, and/or continued squalor, disease, and low life expectancies.

Human beings are a logical species: We take our ideas where they lead. Once we accept the fundamental premise of Owen's piece—that human prosperity harms the planet—the misanthropic ideology of Deep Ecology along with its radical human depopulation agenda becomes a logical next step.

### Agenda: Shutting Down the West's Economies.

I have my differences with *Reason's* science writer, Ron Bailey. But he sure nailed the nub of the problem when reporting on the pro-poverty advocacy he observed at the 2011 climate change conference, aptly entitled "Delusional in Durban":

> Consider some of the proposed cuts in emissions that are being demanded of developed countries. One of the more moderate proposal demands—with proposed phrases in brackets—that "developed countries as a group should reduce their greenhouse gas emissions... [by][at least][40][45][50] per cent from 1990 levels by 2020."
>
> To make it simple, let's take a look at just how a 50 percent cut in U.S. emissions of carbon dioxide might be achieved. Carbon dioxide is the chief greenhouse gas released by burning fossil fuels like oil, natural gas, and coal. According to the Environmental Protection Agency, the U.S. burned enough fossil fuels in 2009 to emit 5.2 billion tons... of carbon dioxide, up from 4.7 billion tons 1990. A 50 percent cut in fossil fuels below 1990 levels would mean cutting annual emissions by roughly 2.8 billion tons in nine years. One way to achieve this would be to shut down completely the 70 percent of America's electric power generation that is fueled by coal and natural gas, plus removing from the roads nearly half of America's 250 million vehicle fleet.[82]

Such cuts would lead to an economic collapse that would make the Great Depression look like the good old days.

### Agenda: Stopping an African Electricity Grid.

I have been to South Africa. It was both a thrilling and, I must say, extremely depressing experience—what I call emotional whiplash. I have never seen such poverty, and yet, the possibilities seem endless if the new South Africa can just grab the brass ring of economic and resource development.

I bring this up because South Africa is the richest sub-Saharan country, and yet the poverty there is bone-crushing; so I can only imagine the human misery found in rest of the continent. And now, global warming hysteria threatens the ability of African nations to get out of the quicksand of destitution by forcing them to refrain from developing their resources and infrastructure.

The *quid pro quo* is that developed nations are to transfer hundreds of billions to the poor countries—in essence, a bribe to keep them from fully exploiting the bounteous resources of the continent. How much of that will actually be used for its stated purposes? We have enough experience with aid to know. Moreover, dependency harms culture. People thrive the more they can be self-reliant.

The best answer for the many tragedies that afflict Africa is development, which leads to stable cultures, smaller families, and the rule of law. That requires, among other things, electrifying the continent. And that means using (mostly) fossil fuels or nuclear power—both anathemas to radical environmentalists. Consider the views of South African nuclear engineer Kelvin Kemm, "Renewables Not the Solution for Africa:"

> So, for us living here in Africa, what perspective must we contemplate? Why should some African children have to do their school homework by candlelight when children in Europe and North America use electric light? African families are entitled to look forward to a modern future for themselves and their descendants. What this implies is that Africa is going to use much more electricity than is being used now. Some African countries are only 5% or 10% electrified. For meaningful economic and social development, they must rapidly double their electricity use, and then double it again, and again.[83]

Who would dispute Kemm? If we care about the welfare and lives of our African brothers and sisters, we must help African countries to develop abundant electricity. But renewables cannot currently do the job , so how will this electricity be produced? Kemm continues:

> Some African countries are endowed with fossil fuels, but most are not. Those countries that do have fossil fuels will use them one way or another. Those that do not will have to find another answer. They will have to find an answer that is realistic and works. Such an answer is compact nuclear reactors that generate 200 MW or less. Wind and solar are not the answer. Wind and solar have their place in the African context but it is not in the large-scale production of baseload electricity.

It is one thing to be a European country and to have wind energy making up, say, 10% of the national electricity mix—it is a totally different story to expect African countries to plan for wind and solar to make up most of the countries' electricity production. These sources are just too variable and intermittent to be a country's prime source of supply.

Africa should be able to do whatever it takes to lift its people into a better way of life. *It doesn't matter whether electrifying Africa would increase or decrease CO2 emissions.* For the sake of *human welfare*, would-be planet savers should stand aside and allow modernity in.

Here's the bottom line: It is anti-human to pursue policies that will keep people living in squalor. In the twenty-first century, everyone should have ready access to electricity and no one should have to heat a hovel by dung-fueled fire. All should have readily available potable water, which also often requires electricity for treatment plants. That takes development, requiring electricity, the generation of which currently necessitates the exploitation of resources, which results in the emission of CO2. Bring it on, say I!

## Authoritarians R Us!

I used to think that people worried about one world government were more than a little paranoid. Today, not so much. Indeed, would-be power grabbers push global warming hysteria, at least in part, as a pretext to impose the beginning of one world government—which would bring technocratic rule by bureaucrats. Indeed, the General Secretary of the United Nations, Ban Ki-moon, hoped that one-world governance could be conceived and gestated toward birth via an anti-climate change treaty agreed to at Copenhagen. From his column in the *New York Times*: "Looking forward to Copenhagen, I have four benchmarks for success: ... A deal must include an equitable global governance structure. All countries must have a voice in how resources are deployed and managed. That is how trust will be built."[84]

The United States remains unlikely (for now) to grant the UN or any other international agency the power to determine how our resources are developed or to manage our economy. But the vaunted "international community" has no such scruples. That means our individual rights depend on maintaining absolute control over our national sovereignty. Participate in international accords, sure. Submit to world governance, no way!

Some (not I) would happily accept benign world governance if it meant the world could be saved. But even conceding for the sake of argument—and only for that purpose—that acceding to international governance could lead to better environmental husbandry, how long would such an all-powerful political structure remain "benign?" Not very. Already, before such an international system has been enacted, we see calls for significant infringement of individual liberty and distinct anti-human tendencies. For example, China extolled its own tyrannical population control policies as a positive—Green, if you will—urging that others follow its lead. From a story in *China Daily*:

> Population and climate change are intertwined but the population issue has remained a blind spot when countries discuss ways to mitigate climate change and slow down global warming, according to Zhao Baige, vice-minister of National Population and Family Planning Commission of China (NPFPC). "Dealing with climate change is not simply an issue of $CO_2$ emission reduction but a comprehensive challenge involving political, economic, social, cultural and ecological issues, and the population concern fits right into the picture," said Zhao, who is a member of the Chinese government delegation.

> Many studies link population growth with emissions and the effect of climate change. "Calculations of the contribution of population growth to emissions growth globally produce a consistent finding that most of past population growth has been responsible for between 40 per cent and 60 percent of emissions growth," so stated by the 2009 State of World Population, released earlier by the UN Population Fund.

> Although China's family planning policy has received criticism over the past three decades, Zhao said that China's population program has made a great historic contribution to the well-being of society. As a result of the family planning policy, China has seen 400 million fewer births, which has resulted in 18 million fewer tons of $CO_2$ emissions a year, Zhao said.[85]

As we discussed in a previous chapter, China's one-child policy has led to forced abortion, female infanticide, eugenic child selection, and other despotic activities. That the Chinese leadership could openly defend forced population control as a pro-Green policy—without fear of serious contradiction or scorn—tells us all we need to know about the hysterics' mindset. Hey, think about the carbon reduction potential of mass euthanasia of the unproductive: Killing could become a whole new industry. Soylent Green is people!

It is one thing when a dictator promotes Green tyranny. It is quite another when such ideas are extolled by free people in the West. Alas, some are so fearful of global warming they would sacrifice our hard-won liberty in the name of holding town the temperature. For example, Diane Francis, writing in the *Financial Post*, argued that the world should emulate the Chinese to keep the planet from melting. From the column by Diane Francis:

> The fix is simple. It's dramatic. And yet the world's leaders don't even have this on their agenda in Copenhagen. Instead there will be photo ops, posturing, optics, blah-blah-blah about climate science and climate fraud, announcements of giant wind farms, then cap-and-trade subsidies. None will work unless a China one-child policy is imposed. Unfortunately, there are powerful opponents. Leaders of the world's big fundamentalist religions preach in favor of procreation and fiercely oppose birth control. And most political leaders in emerging economies perpetuate a disastrous Catch-22: Many children (i.e. sons) stave off hardship in the absence of a social safety net or economic development, which, in turn, prevents protections or development.
>
> China has proven that birth restriction is smart policy. Its middle class grows, all its citizens have housing, health care, education and food, and the one out of five human beings who live there are not overpopulating the planet.[86]

*New York Times* columnist Thomas Friedman similarly extolled China's "one-party autocracy" in a 2009 column:

> One-party autocracy certainly has its drawbacks. But when it is led by a reasonably enlightened group of people, as China is today, it can also have

great advantages. That one party can just impose the politically difficult but critically important policies needed to move a society forward in the 21st century. It is not an accident that China is committed to overtaking us in electric cars, solar power, energy efficiency, batteries, nuclear power and wind power. China's leaders understand that in a world of exploding populations and rising emerging-market middle classes, demand for clean power and energy efficiency is going to soar. Beijing wants to make sure that it owns that industry and is ordering the policies to do that, including boosting gasoline prices, from the top down.[87]

How swell of them. Note Friedman wrote *not one word* decrying the China's appalling record of cruel human rights violations. China—*an unmitigated tyranny*—has become, among the hysterics, a shining example to emulate.

## Punish the Cookers of the Planet

The more the global warming debate rages, the more it seems political rather than scientific, with dissenters branded heretics—even criminals—by the true believers. NASA's former honcho, James Hansen, is one of the field's foremost want-to-be thought policemen. Hansen has seriously called for the jailing of oil executives for committing "crimes against nature" for being global warming "deniers":

> James Hansen, one of the world's leading climate scientists, will today call for the chief executives of large fossil fuel companies to be put on trial for high crimes against humanity and nature, accusing them of actively spreading doubt about global warming in the same way that tobacco companies blurred the links between smoking and cancer.[88]

There's no such thing—*yet, see Chapter 6*—as a "high crime" against nature. Moreover, international crimes against humanity are defined by treaties. But don't expect legal technicalities to stop Hansen's hysteria.

Such personalities are precisely the kind of people who should be kept far away from power. In 2008, Hansen testified in favor of the vandalism of a coal-fired power plant that was admittedly committed by UK supporters of Greenpeace as a means of stopping global warming. Hansen took the stand in support of the claim that trespassers had a "lawful excuse" in causing close to $60,000 in damage to the plant: They were protecting the planet from the coal burning involved in the generation of the plant's electricity! Alas, his testimony may have resonated with the jury, as the perpetrators were found not guilty of the very acts they admitted to committing.[89]

Global warming hysteria not only allows its sufferers to rationalize criminal activity to cool the planet, but it leads some to advocate the criminalizing of speech about the issue with which they disagree. Thus Donald A. Brown, University of Pennsylvania professor of something called "Environmental Ethics, Science, and Law," wants "disinformation"—e.g., opposition to the warming political agenda—considered a crime against humanity. First, Brown jumps the shark about the coming warming-caused destruction:

> These harms include deaths[,] injuries, hunger, and disease from droughts, floods, heat, storm-related damages, rising oceans, heat impacts on agriculture, loss of animals that are dependent upon for substance purposes, social disputes caused by diminishing resources, sickness from a variety of diseases, the inability to rely upon traditional sources of food, the inability to use property that people depend upon to conduct their life including houses or sleds in cold places, the destruction of water supplies, and the inability to live where [sic] has lived to sustain life. In fact, the very existence of some small island nations is threatened by climate change[.][90]

With so much allegedly at stake, only a criminal sanction will prevent "deniers" from confusing the public and interfering with needed policy agendas:

> As long as there is any chance that climate change could create this type of destruction, even assuming, for the sake of argument, that these harms are not yet fully proven, disinformation about the state of climate change science is extraordinarily morally reprehensible if it leads to non-action in reducing climate change's threat when action is indispensable to preventing harm.

And here is Brown's way-over-the-top kicker:

> Clearly this is a new type of crime against humanity. Skepticism in science is not bad, but skeptics must play by the rules of science including publishing their conclusions in peer-reviewed scientific journals and not make claims that are not substantiated by the peer-reviewed literature. The need

for responsible skepticism is particularly urgent if misinformation from skeptics could lead to great harm. For this reason, this disinformation campaign being funded by some American corporations is some kind of new crime against humanity.

The international community does not have a word for this type of crime yet, but the international community should find a way of classifying extraordinarily irresponsible scientific claims that could lead to mass suffering as some type of crime against humanity.

Good grief. The Holocaust was a crime against humanity. American slavery was a crime against humanity. The gulags were a crime against humanity. The Rape of Nanking was a crime against humanity. Anyone who can't distinguish between true evil—*and global warming skepticism*—has no business being a professor at a world-class university.

## WHO NEEDS FREEDOM OF SPEECH?

PAUL THORNTON, AN EDITOR AT THE *LOS ANGELES TIMES*, ANNOUNCED recently that his paper will henceforth refuse to print reader correspondence questioning global warming. His justification for quashing disfavored opinions? "I must rely on the experts—in other words, those scientists with advanced degrees who undertake tedious research and rigorous peer review." And since most support the hypothesis that we are warming the planet, opinions to the contrary are "factual inaccuracy" from "climate deniers"—a pejorative term meant to equate global warming skeptics to those who deny the historical fact of the Holocaust.[91]

Never mind that the "peer reviewed" science shows no statistically significant global warming in 17 years—contrary to the predictions of computer models upon which much of the global warming sector relies for its predictions of coming doom. And never mind that there are reputable climate and other scientists who do not accept the Establishment view about man-made global warming or the proposed solutions to the purported crisis. The *LA Times* has *taken a side* and only readers who agree will be allowed to opine in the published letters.

I wish I could say I was surprised by the *LA Times*'s censorship decision. For the last ten years I have noticed a growing trend in the mainstream media—which claims the mantle of objectivity—to only (or primarily) present the views of "their" side in contentious societal debates and/or to castigate those with heterodox views with pejoratives such a "climate change denier."

A few weeks after the *LA Times*'s censorship move, the moderator of the science thread on the Reddit website also decided to ban "deniers" from commenting. Sayeth Nathan Allen:

> Like our commenters, professional climate change deniers have an outsized influence in the media and the public. And like our commenters, their rejection of climate science is not based on an accurate understanding of the science but on political preferences and personality. As moderators responsible for what millions of people see, we felt that to allow a handful of commenters to so purposefully mislead our audience was simply immoral.

Incongruously claiming a "passionate" dedication to "free speech," Allen nevertheless urged newspapers and other media to follow Reddit's example and ban global warming skeptics from opining.[92]

Apparently the global warmists now believe their case to be so weak—and apparently their readers so gullible—that they feel the need to squelch differing opinions. That isn't discourse. It is propaganda.

## Perhaps the Fever is Breaking

And yet… with global warming's unexpected nearly two-decade "pause" in over-heating the planet, rationality may finally have penetrated some of the meme's prime proponents. For example, James Lovelock, the radical environmentalist who came up with the idea that the Earth is a living organism and who once claimed global warming was our doom, has now poured cold water on climate change fever. From MSNBC:

> James Lovelock, the maverick scientist who became a guru to the environmental movement with his "Gaia" theory of the Earth as a single organism, has admitted to being "alarmist" about climate change and says other environmental commentators, such as Al Gore, were too.

Lovelock, 92, is writing a new book in which he will say climate change is still happening, but not as quickly as he once feared.

He previously painted some of the direst visions of the effects of climate change. In 2006, in an article in the U.K.'s *Independent* newspaper, he wrote that "before this century is over billions of us will die and the few breeding pairs of people that survive will be in the Arctic where the climate remains tolerable."

However, the professor admitted in a telephone interview with msnbc.com that he now thinks he had been "extrapolating too far."[93]

Ya think?

The new book will discuss how humanity can change the way it acts in order to help regulate the Earth's natural systems, performing a role similar to the harmonious one played by plants when they absorb carbon dioxide and produce oxygen. It will also reflect his new opinion that global warming has not occurred as he had expected. "The problem is we don't know what the climate is doing. We thought we knew 20 years ago. That led to some alarmist books—mine included—because it looked clear-cut, but it hasn't happened," Lovelock said.

"The climate is doing its usual tricks. There's nothing much really happening yet. We were supposed to be halfway toward a frying world now," he said. "The world has not warmed up very much since the millennium. Twelve years is a reasonable time… it (the temperature) has stayed almost constant, whereas it should have been rising—carbon dioxide is rising, no question about that," he added. He pointed to Gore's "An Inconvenient Truth" and Tim Flannery's "The Weather Makers" as other examples of "alarmist" forecasts of the future.

I repeat: Ya think? Well, at least Lovelock understands that the meltdown has not proceeded as predicted and has the intellectual flexibility to adjust his thinking accordingly. Other hysterics should take a cue. They might regain some of their lost credibility.

Global warming hysteria is the most visible environmental issue of the current age. But it isn't nearly the most radical. We now turn our focus to an even more misanthropic—and potentially harmful—radical environmental agenda, known as the "rights of nature."

# 5. The "Rights" of Nature

"The environment should compete with religion," a 2007 position paper published by the Institutional Institute for Sustainable Development argued, as faith is "the only compelling, value-based narrative available to humanity."[94]

The Institute advises international bureaucrats working for the United Nations Environmental Program (UNEP) about how to effectively persuade the world to follow UNEP's recommended policies. But how do you start a new Earth religion in a secular age in which religious appeals have lost much of their resonance?

Establish what could be described as a materialistic theocracy.

Okay. How to do that? Personalize "nature", and grant "her" human-like "rights." Once we see nature as us, we will live gently on the Earth—or else!

## Deconstructing the Meaning of "Rights"

Granting "rights" to nature distorts the very meaning of the term. Generally stated, rights are properly understood as moral entitlements embodied in law to protect the lives and liberty of all people. Rights are not earned: They come with the package of being a member of the human race. This principle was most eloquently and succinctly enunciated in the Declaration of Independence with the assertion of the self-evident truth that "all men" (and women) are "created equal" and are therefore "endowed by their Creator with certain unalienable Rights," among which "are Life, Liberty and the pursuit of Happiness."

For hundreds of years, implementing Jefferson's timeless assertion of the unique importance of man—a concept sometimes called human exceptionalism—was the noble goal of Western Civilization. Toward this end, invidious bigotry has been repeatedly and courageously assailed, and one by one, religious persecution, slavery, Jim Crow, exclusive male suffrage, etc., have either collapsed altogether or been materially weakened on their way toward hoped-for and unlamented extinction.

But then, as we finally approached the point of triumph, many among the intellectual elite veered sharply away from the embrace of human exceptionalism and began to deny the unique moral worth and importance of humankind.

In bioethics, utilitarian philosopher/advocates such as the late Joseph Fletcher (who also gave us situational ethics) and Princeton's Peter Singer (the world's foremost proponent of infanticide) asserted that being human is irrelevant to possessing moral value. Full protection must be earned by possessing cognitive capacities—such as being able to value your own life—with those not measuring up denigrated as so-called human "non-persons" who lack the rights to life and bodily integrity.[95]

As the bioethics movement dehumanized the most weak and vulnerable among us, the animal rights movement—as distinct from animal welfare advocacy—agitated against rights being thought of as exclusively human. Animals possess equal rights to humans, movement activists argue, because the ability to suffer is the quality that conveys value. Since both cows and humans experience pain, they are equal—meaning dairy farming and cattle ranching are as odious as human slavery.

PETA's Ingrid Newkirk put it this way: Humans should not have special rights because "A rat, is a pig, is a dog, is a boy."[96] Such moral relativism leads to outrages such as PETA's infamous "Holocaust on Your Plate" pro-vegan campaign, which asserted, "The leather sofa and handbag are the modern equivalent of the lampshades made from the skins of the people killed in the death camps."[97] (PETA later apologized for "an insensitive comparison."[98])

Like a spreading stain, we now see this same deconstruction of rights and human exceptionalism emanating out of the environmental movement—with the potential to do even greater harm.

## The Universal Declaration of the Rights of Nature

Every radical movement needs a manifesto. So too, the movement for the rights of nature. It is as if they think Jefferson wrote, "We hold these truths to be self-evident, that all flora and fauna are created equal, that mountains are endowed by their creator with certain unalienable rights."

The nature rights movement is both anti-rational and anti-human, and it explicitly promotes the idea propounded by Deep Ecology and Gaia Theory that the Earth is a living entity. This view has been embraced by many influential environmental activists, including President Barack Obama's former "green czar" and current *CNN Crossfire* host, Van Jones.[99] It is expressly stated in the "Universal Declaration of Rights of Mother Earth," which proclaims: "Mother Earth is a living being" in relationship with "all other beings," an assertion that would apply to everything on the planet—from us to viruses and stink weeds. The Declaration also declares that Mother Earth's rights are "inalienable" because they "arise from the same [unidentified] source of existence." No mention of God, but I guess if you are trying to establish a new quasi-religion you don't want to boost the competition.[100]

Granting rights to nature would mean that nature's various elements and organisms cannot be discriminated against. Accordingly, the Declaration solemnly asserts: "Mother Earth and all beings are entitled to all the inherent rights recognized in this Declaration *without distinction of any kind, such as may be made between organic and inorganic beings[!], species, origin, use to human beings, or any other status*" [Emphasis added.].

So, you and a fly—heck, you and an outcropping of granite—are equal. And if there are "conflicts" among the organic and inorganic rights-bearers, well, they "must be resolved in a way that maintains the integrity, balance and health of Mother Earth"—whatever that means. This raises the next question: What exactly are these putative "inherent rights of Mother Earth?" Well, according to the Declaration, they include:

+ "The right to life and to exist." Ironically, many supporters of nature rights likely do not favor the right to life for all humans, particularly those not yet born; yet they want to extend the right to mosquitoes and fungi.

+ "The right to be respected." Don't worry, pond scum, I respect your algae essence!

+ "The right to regenerate its bio-capacity and to continue its vital cycles and processes free from human disruptions." Do you detect the inherent anti-humanism? If ever enforced, this "right" would put an end to human thriving and enterprise.

+ "The right to water as a source of life." But what about the right of water not to be consumed? After all, water is part of nature too!

So, does the "right to life and to exist" put the lion and shark out of business? Of course not. "Every being has the right to wellbeing and to live free from torture or cruel treatment *by human beings*." [Emphasis added.] We have no commensurate rights against other rights-holders or Mother Earth, nor do they have such rights *vis-à-vis* each other.

Why the distinction between us and the rest of nature? First, it is a tacit admission of human exceptionalism. But more to the point, these discussions aren't really about "rights" at all. Rather, they seek to impose extreme and self-sacrificing duties on human beings as an ideological dogma or a quasi-pagan sacrament.

Consequently, the Declaration concludes by pronouncing the "Obligations of Human Beings to Mother Earth," which include a "guarantee that the damages caused by human violations of the inherent rights... are rectified and that those responsible are held accountable for restoring the integrity and health of Mother Earth," and the empowerment of "human beings and institutions to defend the rights of Mother Earth and of all beings." Talk about a full employment act for lawyers!

## THE "NATURE RIGHTS" MOVEMENT IS GAINING STEAM

UNFORTUNATELY, THE MOVEMENT TO GRANT RIGHTS TO "MOTHER Earth" is growing, not dissipating, thanks in large part to a radical American environmental group called the Community Environmental Legal Defense Fund (CELDF). According to a CELDF press release, the idea behind nature rights is to "change the status of ecosystems from being regarded as *property* under the law to being recognized as *rights-bearing entities.*"

CELDF's biggest success to date was the adoption of nature rights into the Constitution of Ecuador. Toward this end, the Ecuador Constitution now reads: "Persons and people have the fundamental rights guaranteed in this Constitution and in the international human rights instruments. Nature is subject to those rights given by this Constitution and Law."

What does this co-equal legal status between humans and nature mean? Article 1 states: "Nature or Pachamama, where life is reproduced and exists, has the right to exist, persist, maintain and regenerate its vital cycles, structure, function and its processes in evolution."[101] It is worth noting at this point that "Pachamama" was the Incan goddess of earth and fertility.[102]

Ecuador's Constitution also empowers any radical environmental organization anywhere in the world to travel to Ecuador and legally enforce nature's fundamental rights. Article 1 continues: "Every person, people, community, or nationality, will be able to demand the recognition of rights before the public bodies [courts, governmental agencies, etc.]."

Ecuador has gone far beyond establishing strict environmental protections as a human duty. Rather, in a stunning act of self-demotion, it has reduced humankind to *merely one among the millions of life forms on Earth*—no more worthy of protection than any other aspect of the natural world. In other words, Ecuador's flora and fauna all now have the constitutional and legally enforceable right to exist, persist, and regenerate their vital cycles.

This is so wildly beyond anything we have seen before that the ultimate impact is incalculable. But the potential harm to human welfare seems virtually unlimited.

Take for example a farmer who wishes to drain a swamp to create more tillable land so as to better support his family. Now, the swamp has co-equal rights with the farmer, as do the mosquitoes, snakes, algae, rats, spiders, trees, fish, etc., that reside therein. And since draining the swamp would unquestionably destroy "nature" and prevent it from "existing, persisting, and regenerating its life cycles," one could imagine the farmer (or for that matter, miners, loggers, fishermen, builders, or other users and developers of land and resources) not only prevented from earning his livelihood, but perhaps even accused of *oppressing* nature—a breathtaking concept! Moreover, if the government failed to protect the rights of the swamp (or the trees, the animals on the mineable mountain, the schools of fish, etc.) the Constitution explicitly allows alien radicals like the Community Environmental Legal Defense Fund to descend on Ecuador and sue to thwart land users and resource developers from deciding what to do with their own land—if indeed land can still be actually owned. The mind simply boggles.

Some might respond that Ecuador is a small, very left-wing country not worthy of too much concern. Perhaps. But the concept is spreading. Bolivia also has adopted the "rights of nature" into law. So too have more than thirty municipalities in the United States, including Pittsburgh and Santa Monica.[103]

The wording of these local laws is virtually identical to that in the Constitution of Ecuador, albeit without reference to Pachamama. Thus, Santa Monica's ordinance requires the law to "recognize the rights of people, natural communities, and ecosystems to exist, regenerate and flourish."

Such local laws are designed primarily to thwart energy corporations from engaging in the controversial hydraulic fracturing ("fracking") method of oil and gas extraction within city limits.[104] But granting rights to nature isn't required to stop fracking. A city that doesn't want it conducted within town limits can pass an ordinance prohibiting it.

The real point—particular in places like Santa Monica—is ideological. The city is a hotbed of very left-wing politics. Granting rights to nature is thus intended to reinforce the "progressive" notion that humans are merely one equal part of nature rather than the one being capable of—at least to a degree—standing apart from it and molding it to suit our thriving.

Santa Monica's law is especially ironic in this regard. I was born in Los Angeles and lived in the area for the first forty-two years of my life. I know Santa Monica: There is no "nature" remaining in the city, unless you include the mackerel that take shelter under the pier. To actually protect the "rights" of indigenous flora and fauna, Santa Monica would have to tear up the huge beach parking lot, destroy the pier, and uproot beautiful Palisades Park so the "natural communities" could reestablish the bluff above the Pacific to its pre-development state.

Right! And watch land values and revenues drop like a stone thrown in the ocean. Never. Going. To. Happen.

And don't expect the mainstream media to sound the alarm about the forces seeking to tear down the unique moral status of human life. For example, the woolly-headed *Los Angeles Times*, while not officially endorsing the concept, called nature rights "intriguing."[105]

Nature rights also has been warmly embraced by some at the highest levels of the international community. No less a luminary than Ban Ki-moon, the General Secretary of the United Nations, supports "nature rights"—a matter of special concern since the "rights of nature" have been proposed for inclusion in an eventual treaty to fight global warming. Specifically, on pages 44–45, the draft proposal reads:

Impacts:...

38. *Ensuring* the full respect of human rights, including the inherent rights of indigenous peoples, women, children, migrants and all vulnerable sectors;

39. [*Recognizing* promoting and defending the rights of nature to guarantee harmony between humanity and nature ensuring the prevalence of all elements of nature over market interests];

40. [*Ensuring* that ecological functions of Mother Earth will not be commodified in order to guarantee the rights of nature].[106]

The bracketed sections have not yet been accepted for inclusion in the proposed treaty. However, this is nothing less than a proposed official adoption of a binding and international Earth-religion on the people of the entire world.[107]

I understand this is draft language that might well not make the final cut. But I believe that many global warming hysterics would enthusiastically embrace the rights of nature, because it would:

+ Undermine human exceptionalism by conflating humans—who exclusively possess rights—with flora, fauna, viruses, dirt, and rocks, all of which are part of nature.

+ Materially impede development and the creation of wealth by placing a huge roadblock in front of economic enterprise, since once nature had rights, the natural world would have to be given equal consideration with humans in determining public policy. Ironically, this would push us toward greater poverty, preventing much wealth redistribution, which is a key goal of global warming hysterics.

Talk about eating your own tail! Whatever the merits or demerits of the warming hypothesis, this "solution" is a catastrophe in the making. Nature doesn't have rights. Humans do. Qualified, to be sure, by a duty to properly husband the world, these rights certainly include the promotion of human prosperity and thriving.

## A PRESCRIPTION FOR UNENDING LITIGATION

HOW WOULD THE "RIGHTS OF NATURE" BE BALANCED WITH THE NEEDS of humans? Why, in court, of course—at least, that is the vision CELDF sells on its website (my emphasis):

**What happens when nature's rights and human rights conflict?**

The same thing that happens when different human rights *conflict—a court weighs the harms to the interests, and then decides how to balance them.* Given that ecosystems and nature provide a life support system for humans, their interests must, at times, override other rights and interests. Otherwise, we wouldn't have a planet to inhabit that would support our continued existence. Of course, humans are an integral part of nature as well, which means that human needs must also be considered when the rights and interests of ecosystems come into conflict with those of humans.[108] [Emphasis added.]

It isn't as if we are suffering from a shortage of environmental litigation! But even the existing legal and regulatory gauntlet that must be run before development can commence would pale in comparison to what would be in store if nature were granted legally enforceable rights.

Here's why: Nature rights laws grant legal standing to anyone who wants to bring suit to protect nature's right to "exist, regenerate, and flourish." Talk about a full employment guarantee for lawyers! Imagine the courtroom backlog that would be created if "nature" could sue (funded by well-heeled radical environmentalist groups) every time enterprising humans wanted to act enterprisingly with their own property. Indeed, imagine trying to obtain a liability insurance policy. Good luck with that! And even if nature rights lawsuits lost, they would increase the cost of development exponentially and create a profound chilling effect—if you will pardon the pun—on all human enterprise.

Properly speaking, nature rights would prevent us from truly owning property. We would become, at best, fiduciaries for all of the life forms on the particular tracts of land that we no longer owned. And here the truth begins to shine. The rights of nature is a concept guaranteed to destroy free markets, thwart capitalistic enterprise, shrink economies, reduce wealth, and depress living standards. In the West, that would create damaging recessionary policies.

Such self-destructive polices would have a particularly pernicious impact in the developing world, where granting equal rights to bushes, snakes, perhaps even rock outcroppings —all parts of nature—would thwart the ability of people to liberate themselves from destitution, leading to shorter and more brutal human lives, and ironically, to higher birthrates, since poverty and large families typically go hand-in-hand.

Any movement that puts the supposed rights of nature above the thriving of people can only be called anti-human. If you find it hard to believe that nature rights will be pushed to its logical conclusion, read the next two chapters.

# 6. Pea Personhood

THE TWISTED THINKING THAT CONJURES NATURE RIGHTS HAS NOW spawned other environmental nonsense notions such as "plant personhood" or "plant dignity." Again, I am not talking about legitimate environmental considerations such as protecting endangered species or shielding threatened ecosystems from destructive despoliation. Rather, I am talking about those who argue that we should assign moral value to individual plants in much the same way we do (or should) to each individual human being.

## Plant Dignity

NONSENSE, YOU SAY? IF ONLY! IN SWITZERLAND, PLANT DIGNITY IS THE law. In the 1990s, the Swiss constitution was amended to guarantee that "account... be taken of the dignity of creation when handling animals, plants and other organisms."[109] There was only one problem: Because the law's passage preceded the moral analysis that could justify it, nobody knew what "dignity of plants" actually meant.

This is a job for "Experts"! So, the Swiss government appointed the big brains of the Swiss Federal Ethics Committee on Non-Human Biotechnology to figure it all out.

The resulting report, *The Dignity of Living Beings with Regard to Plants* is enough to short-circuit the brain. First, it determined that the arbitrary killing of flora is morally wrong.[110] Why? A "clear majority" of the panel adopted what it called a "biocentric" moral view, meaning that "living organisms should be considered morally for their own sake because they are alive." Thus, the panel determined that we cannot claim "absolute ownership" over plants and, moreover, that "individual plants have an inherent worth."

Moreover, because we can't morally own plants, and due to their supposed intrinsic dignity, "we may not use them just as we please, even if the plant community is not in danger, or if our actions do not endanger the species, or if we are not acting arbitrarily." "Plant community"? Now those are two words I never expected to see joined together.

The committee offered one illustration of "arbitrary harm": A farmer mows his field (apparently an acceptable action, perhaps because the hay is intended to feed the farmer's herd; the report doesn't say). But then, while walking home, he casually "decapitates"—yes, that is the description—some wildflowers with his scythe. The panel unanimously decried this act as immoral, though its members could not agree exactly why. The report states, opaquely: "At this point it remains unclear whether this action is condemned because it expresses a particular moral stance of the farmer toward other organisms or because something bad is being done to the flowers themselves."

This much *is* clear: Switzerland's enshrining of "plant dignity" is a consequence of the misanthropic cultural epidemic afflicting Western civilization, among the symptoms of which are an inability to think critically and to distinguish serious from frivolous ethical concerns. It also reflects the triumph of a radical anthropomorphism that envisions elements of the natural world as morally equivalent to that of people.

This sad circumstance calls to mind a quip I have sometimes used during debates about human exceptionalism: When my opponent denigrates species distinctions as a "fiction"—which they base on life evolving out of the same primordial ooze and humans sharing many genes with other life forms—I usually respond jokingly: "Well, if you want to get reductionist, let's go all the way. Carrots are made out of carbon-based molecules. So are people. That means there is no difference between us and carrots!"

Now, think about my rejoinder as we ponder the Swiss committee's reasoning for concluding that individual plants may be "part of the moral community." In ethics, a "moral community" is commonly defined as "all those beings that one holds in moral regard. i.e., those beings that you need to think 'but is this right' before you do something that could affect them."[111] So those on the Swiss committee who favored including plants as part of our "moral community" basically wanted people to ask "But is this right?" before weeding their gardens!

To be fair, the Swiss committee was badly splintered about this question. From the report: "Some members were of the opinion that plants are not part of the moral community, because they do not satisfy the conditions for belonging to this community." For example, some claimed that plants aren't "sentient," and thus do not warrant such moral concern. (More on this below.)

Some committee members worried about the impact of granting moral community status to individual plants: "Others argued that plants should not belong to it, because otherwise human life would be morally over-regulated." Apparently, these folk didn't get the memo: Radical environmentalism craves over-regulation.

Another group still asserted that there are "particular situations in which people should refrain from something for the sake of a plant unless there are sufficient grounds to the contrary." Why?

> This opinion was justified either by arguing that plants strive after something [!!!!], which should not be blocked without good reason, or that recent findings in natural science, *such as the many commonalities between plants, animals and humans at molecular and cellular level,* remove the reasons for excluding plants in principle from the moral community. [Emphasis added.]

In other words, my joke was accepted as a serious moral argument by members of the Swiss committee!

Wait; it gets worse. The majority of the committee believed that plants are sentient, or at least, refused to say that they are not:

> The majority of the committee members at least do not rule out the possibility that plants are sentient, and that this is morally relevant. A minority of these members considers it probable that plants are sentient. Another minority assumes that the necessary conditions for the possibility of sentience are present in plants. The presence of these necessary conditions for sentience is considered to be morally relevant.

Plants are living beings. But sentience means the ability to be "responsive to or conscious of sense impressions."[112] A fly senses the swatter descending and escapes being squashed. In contrast, plants react to their environment; for example, flowers point toward the sun. But those are chemical reactions. Plants are not "aware" of what they are doing in even the same rudimentary way as the fly. They are not sentient and cannot by their natures be sentient.

Finally—and this is very telling about where we are as a culture—the Swiss ethicists considered and rejected "theocentrism" (being part of God's creation as the root of dignity), "ratiocentrism" (the capacity to reason as the root of mattering for their own sake), and "pathocentrism" (sentience as the basis for moral worth—an animal rights ideology). But the committee did not even consider "humancentrism,"—i.e., human exceptionalism—the principle that being human matters the most morally regardless of the value we convey to other life forms on the planet.

Plant dignity, if widely accepted, could cause great harm to humanity—and not just in a distorted view of human value—a matter of great import in and of itself. Rather, by including plants in the moral community, we could eschew taking actions of great benefit to us "for the plants."

Such foolishness may have already started. As described in a column by the award-winning *Wall Street Journal* reporter Guatam Naik, a molecular biologist at the University of Zürich named Dr. Beat Keller wanted Swiss government permission to perform a field trial on genetically modified wheat, which he hoped would be resistant to fungus. From the story: "He first had to debate the finer points of plant dignity with university ethicists. Then, in a written application to the government, he tried to explain why the planned trial wouldn't 'disturb the vital functions or lifestyle' of the plants. He eventually got the green light."[113]

Some might say, "So, Keller had to jump through a few extra hoops to get the go-ahead: So what?" First, consider the hunger that could be ameliorated if wheat could be fashioned so that it could be stored for longer periods. The human benefit alone would be tremendous. But because the wheat has intrinsic dignity under Swiss law, *that didn't matter.* If that doesn't raise alarms about the danger to human welfare and the impingement upon human freedom posed by the nature and plant rights movements, what will?

Requiring Keller to assure the ethicists that the wheat wouldn't be harmed by the experiment reflects a wildfire of misanthropy flaring throughout the world, as well as a naive and irrational romanticizing of nature that tends to elevate animals and plants to the status of people, which is to say, to deflate our self-perception as a species.

And that leads to this bitter irony: Rather than arguing that his proposed experiment could benefit humankind, the scientist was forced to plead that he be allowed to proceed because his work would *benefit the wheat!* Again, from Naik's reporting (my emphasis):

> When applying for a larger field trial, he ran into the thorny question of plant dignity. Plants don't have a nervous system and probably can't feel pain, but no one knows for sure. So Dr. Keller argued that by protecting wheat from fungus *he was actually helping the plant*, not violating its dignity—and helping society in the process.

A little fact of biology seems to have escaped the funding committee's attention. The wheat that would be "helped" by the genetic modification would be grain seeds harvested from a plant killed in the reaping. The grain itself is either dead, or a viable seed. So what now, "seed dignity"? Ow! My head hurts!

And here's another consideration: Behind the lost-in-their-own-brains intelligentsia, lurk thousands—perhaps more—zealots who have hyper-romanticized nature to the point that they act out against those seeking to use science to help humans. Thus, Dr. Keller's opponents were not limited to government bureaucrats:

> In June [2008], about 35 members of a group opposed to the genetic modification of crops, invaded the test field. Clad in white overalls and masks, they scythed and trampled the plants, causing plenty of damage.
>
> "They just cut them," says Dr. Keller, gesturing to wheat stumps left in the field. "Where's the dignity in that?"

What folly. We live in a time of cornucopian abundance and plenty, yet countless human beings are malnourished, even starving. In the face of this cruel paradox, wringing our hands about the purported dignity of plants is the true immorality.

This kind of folly is entirely predictable when nature becomes an object of quasi-worship! It is what happens when humans so degrade the worth of their own species that they consider themselves to be merely another aspect of the natural world, neither more nor less important than other fauna and flora. When one equal species impinges on the "rights" of another, action must be taken!

## Peas Are Persons, Too

The Swiss plant dignity virus appears to be catching. The April 28, 2012 *New York Times* ran a piece in its Sunday opinion section by a university professor—*of course!*—claiming that it is *unethical to eat certain plants*.[114] According to Michael Marder,[115] recent discoveries show that peas communicate with each other through their root systems and soil.

Needless to say, peas being plants, pea "communication" doesn't involve the least level of sentience, not to mention the application of rationality. It is a purely chemical response to environmental stimuli. Should mere chemical communication elevate the moral value of peas? Yes, according to Marder:

> When it comes to a plant, *it turns out to be not only a what but also a who—* an agent in its milieu, *with its own intrinsic value or version of the good.* Inquiring into justifications for consuming *vegetal beings* thus reconceived, we reach one of the final frontiers of dietary ethics. [Emphasis added.]

Good grief. Using the word "who" imputes personhood. And plants don't have any "version of the good"—or for that matter, the bad: They are plants!

Marder then claims that plant sophistication means we should not eat vegetation unless the plants live for several growing seasons:

> The "renewable" aspects of perennial plants may be accepted by humans as a gift of vegetal being and integrated into their diets. *But it would be harder*

*to justify the cultivation of peas and other* annual *plants*, the entire being of which humans devote to externally imposed ends. [Emphasis added.]

People are starving in the world, *and Marder worries about the ethics of eating peas and carrots!!* Worse, his piece runs in the Sunday opinion section of the nation's "paper of record"! Marder's thesis was even explored respectfully on National Public Radio![116] (Yes, I'm yelling.)

If Marder's piece—and subsequent book—were just a bizarre outlier, his views might be dismissed with a chuckle and an eye roll. Alas, the plants-are-persons-too meme has been gaining traction in recent years. For example, back in 2009, Natalie Angier, a science columnist for (yes, again) the *New York Times*, marveled, like Marder, about the sophistication of plant biology, and then jumped her own shark by arguing that plants are the most ethical life forms on the planet. From, "Sorry Vegans, Brussels Sprouts Like to Live, Too":

> But before we cede the entire moral penthouse to "committed vegetarians" and "strong ethical vegans," we might consider that *plants no more aspire to being stir-fried in a wok* than a hog aspires to being peppercorn-studded in my Christmas clay pot. This is not meant as a trite argument or a chuckled aside. *Plants are lively and seek to keep it that way.*[117] [Emphasis added.]

I will try to keep my voice down: Surely as a science writer, Angier must know that plants don't "aspire" to anything. But that doesn't stop Angier from larding on the anthropomorphism:

> Just because we humans can't hear them doesn't mean plants don't howl. Some of the compounds that plants generate in response to insect mastication—their feedback, you might say—are volatile chemicals that serve as cries for help. Such airborne alarm calls have been shown to attract both large predatory insects like dragon flies, which delight in caterpillar meat, and tiny parasitic insects, which can infect a caterpillar and destroy it from within.

Please. It's merely natural selection in action, not a cry for help. And get this ending: "It's a small daily tragedy that we animals must kill to stay alive. Plants are the ethical autotrophs here, the ones that wrest their meals from the sun. Don't expect them to boast: they're too busy fighting to survive." Good grief. Plants are not "ethical." That requires thought, a sense of right and wrong, and the free will to choose between them. Besides, Venus fly-traps digest insects alive. Oh, the immorality!

Why is this happening? Our accelerating rejection of the Judeo-Christian worldview, which upholds the unique dignity and moral worth of human beings, is driving us crazy. Once we knocked our species off the pedestal of exceptionalism, it was only logical that we would come to see fauna and flora as entitled to rights.

Eschewing humans as the pinnacle of "creation" (to borrow the term used in the Swiss constitution) has, over the last several decades, caused environmentalism to mutate from conservationism—a concern to properly steward resources and protect pristine environs and endangered species— into a willingness to thwart human flourishing in the name of valuing the dignity of plants or to otherwise "save the planet."

Alas, the craziness isn't about to stop. Indeed, there is an even more radical proposal on the horizon: Some now want to transform resource and land development into a crime akin to the Holocaust. Hold on to your hat; you are about to enter the mad, mad world of "ecocide."

# 7. Ecocide—A Crime
# Against Humanity?

NATURE RIGHTS CAN BE PERCEIVED AS A "SHIELD" SEEKING TO "PRO-tect" the Earth from being developed. As any Roman gladiator would have told us, a shield is scant protection without a sword. So too in radical environmental anti-humanism.

What could become the green sword in the war against humans? How about a law that punishes large-scale development as a heinous felony? If you have read this far in the book, you know better by now than to think "No way." And indeed, such laws are already being advocated all around the world.

## Here Comes "Ecocide"

LITERALLY TRANSLATED, ECOCIDE MEANS KILLING THE ENVIRONMENT, by which advocates actually mean *Earth murder*. CEOs, government officials, and land-owners are all potential defendants who would be arrested and put in the dock of the International Criminal Court alongside international human rights criminals such as Serbia's Radovan Karadzic and Liberia's Charles Taylor. Talk about placing boulders in the road to prosperity.

It is important to note that ecocide wouldn't primarily punish *polluters*—although it would include such despoiling within its jurisdiction. Rather, practically any human enterprise that radical environmentalists loathe would potentially qualify as a heinous "crime against peace."

Let's dig into the details of the proposed international felony. As defined on the Eradicating Ecocide website, "Ecocide is the extensive damage to, destruction of or loss of ecosystem(s) of a given territory, whether by human agency or by other causes, to such an extent that *peaceful enjoyment by the inhabitants* of that territory has been or will be severely diminished." [Emphasis added.][118]

Note that the term "inhabitants" isn't a reference only to human beings. Rather, it includes all life—everything from grass, fish, and insects to mice, snakes, and people. And the diminishment of "peaceful enjoyment" referred to would not require actual pollution, but as we shall see, could mean a declining supply of forage or a loss of foliage caused by almost any use of the land.

In fact, the "crime" of ecocide would be so encompassing that any company involved in large-scale resource development would almost certainly commit it. Not only that, but as I describe below, government regulators that granted permits allowing such development would be deemed co-conspirators in murdering the Earth. According to the "This Is Ecocide" website in 2010:

> **Ecocide arises out of human intervention.** Heavy extraction, toxic dumping, release of pollutants can all result in ecocide...
>
> **Ecocide is a crime of consequence**[.] e.g[.,] where an energy company procures its energy by extracting fossil fuel, as opposed to creation from renewable energy, that would result in ecocide.
>
> **Ecocide is not a crime of intent.** The intention is rarely to render damage on a given territory, more often it is an outcome of another primary (economic or war) activity.[119]

Note that the "This Is Ecocide" website did not cite deliberate ecological despoliation among its examples of ecocide; it made no mention, for instance, of Saddam Hussein's releasing oil into the Persian Gulf and setting oil wells aflame during the first Gulf War. Primarily, ecocide seeks to thwart human enterprise. In fact, there need not even be harm to any living organism: The proliferation of space junk is listed as ecocide.

Dig deeper into ecocide advocacy and the typical anti-free market ideology that drives too much of environmentalism today comes clearly into focus. Thus, a YouTube video, "Ecocide: A Crime Against Peace," states:

> We have come to accept that extraction of natural resources is normal. Just because it is normal does not mean that it is right. Two hundred years ago

companies plundered for profit. Then it was called colonization. Today it is called business.

Back then, extraction often led to conflict. Sometimes it led to war. Now a century of "resource wars" is predicted. The battle to control oil and water has already started. Now natural resources are becoming the reason for war. Unless we change. Do you see what is happening here?[120]

PowerPoint style, the screen then slowly rolls out the phrase "Extraction = Ecocide > Resource Depletion > War," which melts into the summary statement "Ecocide > War."

That's the agenda. But who could be held culpable for the crime? Just about anyone with decision-making authority, whether in government or business. Thus, the Preamble to the Draft Ecocide Act states that the "burden of responsibility... rests jointly with any person or persons, government or government department, corporation or organization exercising a position of superior responsibility" over the now-criminalized enterprise.[121] Worse, the scope of the proposed law is so sweeping that virtually any human activity could theoretically come within its ambit. Again, from the Draft Ecocide Act:

### 2. Risk of Ecocide

Ecocide is where there is a potential consequence to any activity whereby extensive damage to, destruction of or loss of ecosystem(s) of a given territory, whether by human agency or by other causes, may occur to such an extent that:—

(1) peaceful enjoyment by the inhabitants of that territory or any other territory will be severely diminished; and[/]or

(2) peaceful enjoyment by the inhabitants of that territory or any other territory may be severely diminished; and[/]or

(3) injury to life will be caused; and[/]or

(4) injury to life may be caused.

Note that the terms "given territory" and "peaceful enjoyment" are defined so broadly they could encompass virtually any tract of land (however small or large) and almost every conceivable human use: "'territory' means any domain, community or area of land, including the people, water and/ or air that is affected by or at risk or possible risk of Ecocide… 'peaceful enjoyment' means the right to peace, health and well-being of all life."

A farmer owns a 1,000-acre meadow. He decides the time has come to clear and plow the land to grow crops. From the perspective of the beetles, birds, snakes, lizards, weeds, flowers, trees, and ants residing in the meadow, could he not be said to be committing ecocide by destroying their peaceful enjoyment of the meadow, which easily fits into the definition of an "area of land?"

Indeed, the farmer could also be criminally culpable when he reaps the wheat with a combine, killing thousands of snakes, mice, birds, insects, etc., since ecocide is defined as a "crime against nature": "A person, company, organisation, partnership, or any other legal entity who causes ecocide under section 1 of this Act and has breached a non-human right to life is guilty of a crime against nature."[122]

Wait: Advocates would also outlaw "cultural ecocide," a term that isn't limited to humans, since it is defined as "destruction to or loss of a community's way of life." Moreover, the term "ecosystem" has been downsized to include virtually any aspect of the natural world, since it is defined as a "biological community of interdependent living organisms and their physical environment." Hence, the flora and fauna within the hypothetical meadow transformed into a farmer's field would qualify as a "community," and "destroying the way of life" of the flora and fauna inhabiting the "territory" as cultural ecocide.

## OUTLAWING ECOCIDE WOULD BE A CRIME AGAINST HUMANITY

ADOPTING AN INTERNATIONAL TREATY MAKING ECOCIDE A FELONY would be profoundly subversive of human thriving. First, equating resource extraction and/or pollution with genocide and ethnic cleansing trivializes true evil by erecting a moral equivalency between horrors such as the slaughter in Rwanda, the killing fields of Cambodia, the gulags, and the

Nazi death camps on the one hand, and mining, harvesting forests, oil drilling, and oil spills on the other.

More fundamentally, an ecocide law would cause unimaginable human suffering. Remember, the movement doesn't seek merely to regulate or constrain its target economic activities—*it seeks to criminalize them.* Such criminalization would lead to the collapse of developed world economies and force those now living in destitution to remain mired in their misery.

Ecocide champions are not hiding their intentions. In 2011, they sponsored a mock ecocide prosecution against two fictional energy company CEOs. This was no minor exercise held in a college classroom. The trial was held in the courtroom of the English Supreme Court.

Nor did the moot litigants strive over a hypothetical *Exxon Valdez* pollution catastrophe. Rather, the "defendants" were accused of the supposedly horrible crime of extracting oil from the Alberta tar sands—even though the "criminal" product would heat, fuel transportation, create wealth, and liberate the West from its dependence on the ever-volatile Middle East.

Needless to say, the CEOs were found guilty as charged.[123] One fictional executive was sentenced to four years in prison, while the other was required to participate in "restorative justice"—that is, paying to restore the attacked territory to its former state.[124] Vividly illustrating the hubris of the ecocide movement, the defendants were confronted by lawyers claiming to represent virtually everyone and everything on the Earth:

> Bannerman [one of the CEOs] also came face to face with representatives of those who had been adversely affected by the tar sands Ecocide: Jess Philimore represented wider humanity, Carine Nadal represented the Earth, Philippa De Boissiere represented future generations, Peter Smith represented birds and Gerald Amos provided a voice for indigenous peoples[125]

Talk about legal malpractice! Apparently, the legal representative for "wider humanity"—i.e., for all of us—failed to argue that criminalizing extracting oil from tar sands would chill large-scale energy production everywhere in the world, which would result in terrible harm to humans, including wild inflation in the cost of heating our homes and transporting food to market. The barrister representing "future humanity" similarly failed to argue that people could die as a direct consequence, and that ecocide laws would impose impossible-to-bear burdens upon our posterity causing immeasurable and avoidable poverty and want. Nor did the lawyer representing "indigenous peoples"—many of whom live in resource-abundant areas—protest that ecocide would doom his clients to permanent poverty by thwarting their ability to develop the wealth on their own land. Because, you see, the land wouldn't be theirs anymore, but owned in common with every other life form in the territory.

Let's think about other potential weapons used in ecocide. How about wind farms? After all, they likely slaughter millions of birds, bats, and insects every year, depriving these "individuals" of the "peaceful enjoyment" of their flight paths, hunting and nesting areas, and migratory "cultures." Under a law of ecocide, a lawyer representing "the birds" could argue that CEOs of wind farms and the government regulators that permit the farms should wind up as defendants at The Hague. You think I kid? Some opponents of wind farms already accuse operators of committing ecocide.[126]

And let's not forget global warming hysteria. Already, ecocide campaigners advocate punishing carbon dioxide and other greenhouse gas emitters as if they were Mengele in the camps. The following exchange comes from an interview by Polish environmentalist Marcin Gerwin with the world's most energetic ecocide campaigner, the English lawyer, Polly Higgins:

**MG:** Could it [an international law of ecocide] help to stop climate change?

**PH:** If you define climate change as the increase in greenhouse gases then this is legislation that will have direct impact. What it will do is actually stop dangerous industrial activity of businesses that are considered the carbon majors—the unconventional tar extraction is the most obvious one of all... that practice will not continue much longer under the law of ecocide. It fundamentally shifts the playing field very fast in a way where the Kyoto Protocol has failed. And this is about where political will has not been able to make the change very fast. But the law of ecocide can.[127]

Think of the industries, travel, and other human activities that can be obliterated by criminalizing the emission of carbon. Oh, joy, oh rapture, *oh poverty*! Oh, the irony. Higgins travels the world on supposed climate-change-causing jets, kept aloft by carbon-based fuel, pitching the ecocide snake oil.

If ecocide laws pass, you can also forget about freedom of speech and thought. Campaigners envision their law as a means for prosecuting global warming "deniers." Thus *The Guardian* reported:

> Supporters of a new ecocide law also believe it could be used to prosecute "climate deniers" who distort science and facts to discourage voters and politicians from taking action to tackle global warming and climate change.

> "Ecocide is in essence the very antithesis of life," says [Polly] Higgins. "It leads to resource depletion, and where there is escalation of resource depletion, war comes chasing behind. Where such destruction arises out of the actions of mankind, ecocide can be regarded as a crime against peace."[128]

Apparently, where greens dare to tread, freedom fails to follow.

## "Ecocide" Isn't Permanent

Like all demagogues, ecocide campaigners are masters of half-truths, which is another way of saying that they lie. Yes, large-scale development disrupts localized environments and can pollute. But even without ecocide and "rights of nature" laws, the harm that is done can be limited—sometimes completely eliminated—through proper regulatory policies and government inspection.

But beyond that point, the ecociders don't tell their gullible foot soldiers that "ecocide"—to use their polemic term—*isn't permanent*. Indeed, once timber has been harvested, ore extracted, coal mined, and oil squeezed out of shale, companies are often required to restore the land to its pre-development state.

Take Ecocide Public Enemy Number One: the Alberta Tar Sands. If you listen only to ecociders and view their photographs, you would assume that after a "given territory" is exploited for its oil, the company moves on leaving nothing but a destroyed moonscape.

Not true. The Province of Alberta requires companies to both remediate and reclaim the land—a process that begins at the project planning stage and only concludes when the land has been restored near to its pre-development state. Remediation is cleanup of pollutants and contaminants to protect future residents from potential harm.[129] Reclamation restores land to a natural-looking state, and includes replacing topsoil, topographical contouring, replanting native flora, and other efforts aimed at erasing the scar that development would otherwise leave behind. Ironically, these activities are some of the same "restorative justice" approaches that the law against ecocide would require as a criminal sanction.

It is noteworthy that Alberta has required tar sand companies to deposit more than 700 hundred million (Canadian) dollars into a reclamation security trust fund, so that even if the companies go broke, money will be there to restore the land. The province also won't allow a company to complete a project until it receives a "Reclamation Certificate" of having restored the land to a proper state.[130]

This isn't an easy process, of course. Extracting tar sands presents difficult cleanup challenges. On the bright side, the problems faced by the industry have paved the way (if you will) for the creation of whole new industries dedicated to developing efficient tar sands cleanup techniques and technology.[131]

Don't misunderstand me: I am not saying that the companies don't need vigorous watchdogs with sharp teeth. Clearly they do. Indeed there are already reports of missed time limits among tar sands companies. But I do contend that the tar sands extraction (and other similarly crucial resource extraction projects) can be sufficiently regulated to allow us to reap a tremendous bounty for humanity, while concomitantly requiring proper cleanup—all without granting "rights" to nature or throwing government and business leaders into prison.

Ecocide remains on the edges of environmental activism—for the moment. But that doesn't mean it won't move quickly toward more mainstream acceptance—just like the other radical ideas discussed in this book. Already, the ecocide movement counts among its supporters the famous primatologist Jane Goodall, leftist movie star Daryl Hannah, and international New Age guru, Deepak Chopra.[132]

Perhaps more significantly, ten countries have passed ecocide laws, although on a far more limited scale, aimed narrowly at punishing catastrophic pollution rather than thwarting development.[133] For example, Article 358 of the Russian Federation Criminal Code stringently defines ecocide as the "massive destruction of fauna and flora, contamination of the atmosphere or water resources," or "other acts capable of causing ecological catastrophe." This is hardly the same as the expansive prohibitions sought by the ecocide activists.

In the 1970s, the values of Deep Ecology were anathema to most. Ten years ago, granting "rights" to nature would have been laughed off as a pipe dream. Indeed, as we have seen, in contemporary society very radical ideas often gain quick acceptance by a ruling elite growing ever more antithetical to human exceptionalism. Tempted as we may be to laugh it off, we should instead recognize that ecocide—at least as it applies to criminalizing development as opposed to mass pollution—poses a potent threat to our collective well-being.

# Conclusion: Old Faithful is not a Person

W E LIVE IN AN IRONIC AGE. SOME PEOPLE ARE BEING STRIPPED OF their personhood, while at the same time, the birds and the bees and the flowers and the trees have been granted legal rights. To use a Gilbert and Sullivan term, it's all topsy-turvy.

As we have seen, the Earth is increasingly viewed as a living entity—the Earth-goddess Pachamama—as much entitled to legal rights as human beings. UN bureaucrats and their NGO supplicants and enablers are particularly attracted to these agendas, as implementing them would increase their power and bring the world's economies under the jurisdiction of an international technocracy. Meanwhile, the ecocide movement seeks to criminalize large-scale land use and development of resources as "earth murder."

Laws enacting these anti-human values have already been enacted. Plants have intrinsic dignity in Switzerland. Nature has "rights" in Santa Monica. Angola, of all places, is on the verge of enacting an ecocide law—increasing the likelihood that its people will be unable to shake off the shackles of destitution.

And now, New Zealand has taken these issues a radical step further—legally declaring that the Whanganui River is a person. Yes, you read the last sentence correctly: A *river* has been granted full rights of "personhood," as an "integrated, living whole" possessing "rights and interests."[134] Good grief.

What an astounding and disturbing development. If the river needs protection, New Zealand should protect it by law, without falsely personalizing the natural world.

It's not as if that approach doesn't work. Yellowstone National Park, one of the great wonders of the natural world, not only has been stringently protected for more than 140 years, but has been managed so as to allow people to enjoy its wonders and stand in awe in the face of the region's magnificent beauty—and all without declaring that Old Faithful Geyser is entitled to the rights of personhood.

## Why Anti-Human Environmentalism Matters

In an era in which fundamental principles too often are subsumed by emotion and the narrative is king, many will shrug and ask what difference it makes whether we protect the environment as a matter of human duty or to enforce the putative "rights" of nature. What counts, some will say, is "saving the Earth."

I disagree with the apocalyptic premise. But that point aside, I certainly agree that humans have the duty to act responsibly toward the environment—and that establishing proper practices is a legitimate function of law. Unlike the deep misanthropes, however, I believe that promoting human welfare is crucial, and indeed, should often be the first (albeit, not only) consideration.

I understand that there can be a dynamic tension between environmental protection and promoting human welfare—and much room for reasonable debate about striking proper balances. Such debates can get bitter at times. But they are a necessary function of promulgating proper public policies in a free society. And to be sure, there are going to be times when human prosperity will be constrained by needed environmental restrictions: Nobody advocates returning to the bad old days of unrestricted land use; nobody wants to allow heavy industries a license to pollute at will; nobody yearns to drive endangered species into extinction.

But when we purport to protect nature by redefining humanity as merely one equal organism among all the others on Earth, we do ourselves great harm. When we assert that flora and fauna—perhaps even canyons, rivers, glaciers, and other geological phenomena—have "rights," we devalue liberal principles arising from the "Laws of Nature and of Nature's God" (to borrow a notable phrase from the Declaration of Independence) in the same way that wild inflation cheapens the worth of currency. Indeed, if a squirrel or mushroom and all other Earthly entities somehow possess rights, the very concept withers.

Beyond that, granting rights to nature is intellectually nonsensical. "Rights" can only be understood in the human context. University of Michigan professor of philosophy Carl Cohen puts it this way: "A *right*... is a valid claim, or potential claim, that may be made by a moral agent, under principles that govern both the claimant and the target of the claim."[135] Since only humans are moral agents, only humans are capable of possessing rights.

David S. Oderberg, a philosophy professor at the University of Reading, describes this two-way-street concept somewhat differently:

> what matters in the having of rights is twofold: a) knowledge; b) freedom. More precisely, a right holder must, first, *know* that he is pursuing a good, and secondly, must be *free* to do so. No one can be under a duty to respect another's right if he cannot *know* what it is he is supposed to respect.[136] [Emphasis in original.]

In sum, for nature to possess rights, it must also be capable of assuming concomitant duties toward others. Thus, if the rights of nature to "exist, regenerate, and flourish" can be enforced against us, we would have to be able to make the same claim against nature, a farcical notion.

Environmentalists who embrace human *unexceptionalism* declare that once we see ourselves as merely one, equal part of nature, we will treat nature the way we would want to be treated. Yet, no other species cares a whit about protecting the environment. For example, a herd of elephants will destroy everything in its path without a moment's thought about the environmental damage they cause or of the "territory's" co-inhabitants they are killing.

That's not a slam against elephants. They are incapable of such concern. We—and we alone—are the species capable of acting against our own immediate interests to protect the environment. And we are the only ones called upon to sacrifice our own thriving to protect the "rights" of all others on the planet. If that doesn't make us exceptional, what does?

This book opened with the young David Suzuki crudely proclaiming, in 1972, that human beings are "maggots" who go about their lives "crawling around eating and defecating all over the environment." We will close the loop with the Establishment (with a capital-E) octogenarian, Sir David Attenborough—famous for hosting BBC's *The Living Planet* and other nature documentaries. In 2013, Attenborough lambasted humans as a "plague on the Earth" in support of radical depopulation. From the *Telegraph* story:

> The television presenter said that humans are threatening their own existence and that of other species by using up the world's resources. He said the only way to save the planet from famine and species extinction is to limit human population growth. "We are a plague on the Earth. It's coming home to roost over the next 50 years or so. It's not just climate change; it's sheer space, places to grow food for this enormous horde. Either we limit our population growth or the natural world will do it for us, and the natural world is doing it for us right now," he told the Radio Times.[137]

Attenborough is a patron of Population Matters—the UK's largest nonprofit organization advocating for human depopulation. A representative of the trust agreed with Attenborough, calling his misanthropic analogy "apt," and stating that we are indeed "like a plague of locusts, which consumes all it sees and then dies off."[138]

When Suzuki first spoke, such human-phobic statements shocked. Today, they are clichés of environmental advocacy.

But hating humanity is dangerous. The radical goals and dreams of the anti-humanists are simply not consistent with human liberty and individual freedom. Granting rights to nature will throttle prosperity. Ecocide laws would prevent resource-rich but materially poor nations from escaping their plight. Voluntary family planning offers great benefits. But actually reducing our numbers would require tyrannical measures. After all, China's brutal one-child policy has only *slowed* the country's population growth. The actual Chinese population has not diminished.[139]

If such blatantly despotic measures have not reduced China's population, what would it take to force the world population to actually shrink? Quasi-genocidal means—which, needless to say, Attenborough and his colleagues at Population Matters would never consider or support.

What about other, less civilized zealots, possessed of the same anti-human ideology that now permeates mainstream environmentalism—but without the scruples, political tolerance, and policy "inhibitions" of the modern genteel liberal? After all, history illustrates the deadly consequences of allowing anti-humanism to drive policy and culture—as we learn from the multi-faceted oppression sparked by the eugenics movement in the first half of the twentieth century.[140]

The triumph of anti-humanism within environmental advocacy threatens a green theocratic tyranny. Like eugenics, the misanthropic agendas discussed in this book are all profoundly Utopian endeavors, meaning that the perceived all-important ends will come eventually to justify coercive means. Indeed, the convergence of human loathing, concentrated Malthusianism, and renewed advocacy for radical wealth redistribution—all of which are now respected views within the environmental movement, and each of which is dangerous in its own right—threatens calamity.

Don't say you weren't warned.

# ENDNOTES

1. "Flashback 1972—David Suzuki: Humans are just 'maggots' that 'defecate all over the environment,'" *Hauntingthelibrary* (May 12, 2012), http://hauntingthelibrary.wordpress.com/2012/05/12/flashback-1972-david-suzuki-humans-are-maggots-that-defecate-all-over-the-environment/.

2. David Suzuki, Interview by Jian Ghomeshi, *Q with Jian Ghomeshi*, CBC Radio, (November 25, 2009), http://www.cbc.ca/q/blog/2009/11/25/what-do-you-think-about-what-david-suzuki-and-al-gore-have-to-say-1/.

3. "Trailer—The Day the Earth Stood Still (1951)," http://youtu.be/51JoEE_znyI.

4. See the description at "The Happening" (2008), http://www.imdb.com/title/tt0949731/. You can view the trailer at http://youtu.be/TIQ21m1Ks08.

5. "Signs" (2002), http://www.imdb.com/title/tt0286106/.

6. "AVATAR: Get Rid of Human Beings Now!" *Movie Guide*, http://www.movieguide.org/reviews/avatar.html.

7. "Ted Kaczynski," *Wikipedia*, http://en.wikipedia.org/wiki/Ted_Kaczynski.

8. Article is still available on McKitrick's website under the title of "Earth Hour: A Dissent," http://www.rossmckitrick.com/uploads/4/8/0/8/4808045/earthhour.pdf.

9. Viv Forbes, "Earth Hour or Blackout Night?" *Carbon Sense Coalition*, press release (March 27, 2010), Scoop, http://www.scoop.co.nz/stories/WO1003/S00544.htm.

10. Daniel T. Willingham, "Trust Me, I'm a Scientist," *Scientific American* (May 5, 2011), http://www.scientificamerican.com/article.cfm?id=trust-me-im-a-scientist.

11. David Shearman, "Climate change, is democracy enough?" *On Line Opinion*, (January 17, 2008), http://www.onlineopinion.com.au/view.asp?article=6878.

12. Craig Offman, "Jail politicians who ignore climate science: Suzuki," *National Post* (February 7, 2008), http://www.nationalpost.com/news/story.html?id=290513.

13. "Malthusianism," *Wikipedia*, http://en.wikipedia.org/wiki/Malthusianism.

14. Arne Naess, "The Deep Ecology Platform," Foundation for Deep Ecology, http://www.deepecology.org/platform.htm.

15. "Our Mission," *Foundation For Deep Ecology*, http://www.deepecology.org/mission.htm.

16. "Overview: Understanding Gaia Theory," http://www.gaiatheory.org/overview/.

17. Albert J. Bergesen, "Our growing sense of eco-alienation is part of a historic pattern / Earth Day helps us to reconnect," *SFGate* (April 23, 2006), http://www.sfgate.com/opinion/article/Our-growing-sense-of-eco-alienation-is-part-of-a-2536779.php#ixzz25S5heEHV.

18. Wesley J. Smith, *A Rat is a Pig is a Dog is a Boy* (New York: Encounter Books, 2010).

19. James Lovelock, *The Revenge of Gaia* (New York: Basic Books, 2006), 154.

20. Anne Barbeau Gardiner, "Human Sacrifice on the Altar of Gaia," *New Oxford Review* (June 2008), http://www.newoxfordreview.org/reviews.jsp?did=0608-gardiner.

21. Paul Watson, "The Beginning of the End for Life as We Know it on Planet Earth? There is a Biocentric Solution," *Sea Shepherd* (May 4, 2004), http://www.seashepherd.org/commentary-and-editorials/2008/10/30/the-beginning-of-the-end-for-life-as-we-know-it-on-planet-earth-340.

22. "Unabomber's Manifesto," http://cyber.eserver.org/unabom.txt.

23. Joe Weisenthal, "Check Out the Crazy, Baby-Hating Demands of the Discovery Channel Hostage Taker," *Business Insider* (September 1, 2010), http://www.businessinsider.com/discovery-channel-hostage-taker-demands-2010-9.

24. Eric R. Pianka, *The Vanishing Book of Life on Earth*, http://www.zo.utexas.edu/courses/THOC/VanishingBook.html.

25. Eric R. Pianka, "The 'Controversy' over Eric Pianka's Speech," http://uts.cc.utexas.edu/~varanus/Controversy.html.

26. Sarah-Kate Templeton, "Two children should be limit, says green guru," *The Sunday Times* (Feb. 1, 2009), http://www.thesundaytimes.co.uk/sto/style/living/article147247.ece. Also see "Malthusianism," *Wikipedia*, http://en.wikipedia.org/wiki/Malthusianism.

27. Rick Pearcey, "Dr. 'Doom' Pianka Speaks," *The Pearcey Report* (April 6, 2006), http://www.pearceyreport.com/archives/2006/04/transcript_dr_d.php/.

28. Susan Greenhalgh, *Cultivating Global Citizens: Population in the Rise of China* (Harvard University Press, 2010).

29. Jonathan Mirsky, "China's Dreams of Superior Children," *The Wall Street Journal* (Jan. 2, 2011), http://online.wsj.com/article/SB10001424052748704735304576057300258180310.html.

30. Voluntary Human Extinction Movement, http://www.vhemt.org/.

31. http://www.vhemt.org/success.htm.

32. "Paul Kingsnorth," *Wikipedia*, http://en.wikipedia.org/wiki/Paul_Kingsnorth.

33. Mark Newton, "Q & A: Paul Kingsnorth, green activism is 'simply a faction of consumer society'," *Ecologist* (September 21, 2011), http://www.theecologist.org/green_green_living/Q_and_A/1062124/qa_paul_kingsnorth.html.

34. "The Dark Mountain Manifesto," *The Dark Mountain Project* (2009), http://dark-mountain.net/about/manifesto/.

35. Ibid.

36. George Monbiot and Paul Kingsnorth, "Is there any point in fighting to stave off industrial apocalypse?" *The Guardian* (August 17, 2009), http://www.guardian.co.uk/commentisfree/cif-green/2009/aug/17/environment-climate-change.

37. Ibid.

38. "Georgia Guidestones," *Wikipedia*, http://en.wikipedia.org/wiki/Georgia_Guidestones.

39. Jill Neimark, "Georgia's Own Doomsday Stonehenge Monument," *Discover Magazine* (September 9, 2013), http://blogs.discovermagazine.com/crux/2013/09/09/georgias-own-doomsday-stonehenge-monument/#.UjCsEsaThc5.

40. Gerald Traufetter, "Stagnating Temperatures: Climatologists Baffled by Global Warming Time-Out," *Spiegel Online* (November 19, 2009), http://www.spiegel.de/international/world/stagnating-temperatures-climatologists-baffled-by-global-warming-time-out-a-662092-2.html.

41. Damian Carrington, "IPCC officials admit mistake over melting Himalayan glaciers," *The Guardian* (January 20, 2010), http://www.guardian.co.uk/environment/2010/jan/20/ipcc-himalayan-glaciers-mistake.

42. Zac Unger, "Are Polar Bears Really Disappearing?" *The Wall Street Journal* (February 8, 2013), http://online.wsj.com/article/SB10001424127887323452204578288343627282034.html; Paul Waldie, "Healthy polar bear count confounds doomsayers," *The Globe and Mail* (April 4, 2012), http://www.theglobeandmail.com/news/national/healthy-polar-bear-count-confounds-doomsayers/article2392523/.

43. "Kilimanjaro's vanishing ice due to tree-felling," *New Scientist* (September 25, 2010), http://www.newscientist.com/article/mg20727794.400-kilimanjaros-vanishing-ice-due-to-treefelling.html.

44. Apolinari Tairo, "Snow slowly building on Mount Kilimanjaro," *eTurbo News* (March 15, 2011), http://www.eturbonews.com/21762/snow-slowly-building-mount-kilimanjaro.

45. James Taylor, "Global Warming Alarmists Flip-Flop On Snowfall," *Forbes* (March 2, 2011), http://www.forbes.com/sites/jamestaylor/2011/03/02/global-warming-alarmists-flip-flop-on-snowfall.

46. Louise Gray, "Copenhagen climate change conference: world 'has 10 years to reverse trends,'" *The Telegraph* (December 9, 2009), http://www.telegraph.co.uk/earth/copenhagen-climate-change-confe/6770111/Copenhagen-climate-change-conference-world-has-10-years-to-reverse-trends.html.

47. "Global warming a time bomb ticking away," *Daily Times* (January 25, 2005), http://www.dailytimes.com.pk/default.asp?page=story_25-1-2005_pg7_49.

48. Robin Mckie, "President 'has four years to save Earth'," *The Guardian* (January 17, 2009), http://www.theguardian.com/environment/2009/jan/18/jim-hansen-obama.

49. Judd Legum, "Al Gore, NYU Law, 9/18/06," *Think Progress* (September 18, 2006), http://thinkprogress.org/gore-nyu/?mobile=nc.

50. David Adam, "UN scientists warn time is running out to tackle global warming," *The Guardian* M(ay 4, 2007), http://www.guardian.co.uk/environment/2007/may/05/climatechange.climatechangeenvironment.

51. Quoted in Georgina Robinson, "Philosophy and carbon emissions: what should you think?" *The Sydney Morning Herald* (November 26, 2009), http://www.smh.com.au/national/philosophy-and-carbon-emissions-what-should-you-think-20091126-jtln.html.

52. Cheryl Jones, "Frank Fenner sees no hope for humans," *The Australian* (June 16, 2010), http://www.theaustralian.com.au/higher-education/frank-fenner-sees-no-hope-for-humans/story-e6frgcjx-1225880091722.

53. Greg Heffer, "Sir Bob Geldof: 'All humans will die before 2030,'" *Daily Star* (October 6, 2013), http://www.dailystar.co.uk/news/latest-news/342876/Sir-Bob-Geldof-All-humans-will-die-before-2030.

54. Murray Wardrop, "Copenhagen climate summit: Al Gore condemned over Arctic ice melting prediction," *The Telegraph* (December 15, 2009), http://www.telegraph.co.uk/earth/copenhagen-climate-change-confe/6815470/Copenhagen-climate-summit-Al-Gore-condemned-over-Arctic-ice-melting-prediction.html.

55. http://www.esa.int/Our_Activities/Observing_the_Earth/CryoSat/Arctic_sea_ice_up_from_record_low.

56. Mica Rosenberg, "Lizards face extinction from global warming: study," *Reuters* (May 13, 2010), http://www.reuters.com/article/2010/05/13/us-climate-lizards-idUSTRE64C4PV20100513.

57. Wesley J. Smith, "Global Warming Hysteria: Now We Know Why Heterodox Scientific Views Have Been Stifled," *National Review* (February 14, 2010), http://nationalreview.com/human-exceptionalism/325235/global-warming-hysteria-now-we-know-why-heterodox-scientific-views-have-.

58. Kate Ravilious, "How green is your pet?" *New Scientist* (October 23, 2009), http://www.newscientist.com/article/mg20427311.600-how-green-is-your-pet.html.

59. Justin Gillis, "Seeing Irene as Harbinger of a Change in Climate," *New York Times* (Aug. 27, 2011), http://www.nytimes.com/2011/08/28/us/28climate.html.

60. Brian McNoldy, "What happened to hurricane season? And why we should keep forecasting it…" *Washington Post* (September 30, 2013), http://www.washingtonpost.com/blogs/capital-weather-gang/wp/2013/09/30/what-happened-to-hurricane-season-and-why-we-should-keep-forecasting-it/.

61. "Today's Quakes Deadlier Than in Past," Associated Press story on CBSNews.com (June 18, 2008), http://www.cbsnews.com/news/todays-quakes-deadlier-than-in-past/.

62. "How to see and READ the AURA: Part 1," *Thiaoouba Prophecy*, http://thiaoouba.com/seeau.htm.

63. See the FAQs at http://thiaoouba.com/faq.htm. For a sampling of Chalko's books visit the Bioresonant bookshop at http://bioresonant.com/bookshop.html.

64. "Global Warming: Can Earth EXPLODE?" *Bioresonant*, http://bioresonant.com/news.htm.

65. Matthew Liao, "Hand-made humans may hold the key to saving the world," *Bendigo Advertiser* (September 30, 2012), http://www.bendigoadvertiser.com.au/story/366863/hand-made-humans-may-hold-the-key-to-saving-the-world/.

66. Daniel Strain, "To Fight Global Warming, Eat Bugs," *Science* (January 4, 2011), http://news.sciencemag.org/sciencenow/2011/01/to-fight-global-warming-eat-bugs.html.

67. Danielle Demetriou, "Japanese told to go to bed an hour early to cut carbon emissions," *The Telegraph* (June 24, 2010), http://www.telegraph.co.uk/earth/environment/climatechange/7851292/Japanese-told-to-go-to-bed-an-hour-early-to-cut-carbon-emissions.html.

68. Phys.org, "Man-made global warming started with ancient hunters: study," Phys.org (June 30, 2010), http://phys.org/news197131477.html.

69. "Tick Global Warming Ad," YouTube, http://youtu.be/BdSTM-U7cTk.

70. "According to Habitat Heroes Study, Children Fear the End of the Earth," *Reuters*, http://www.reuters.com/article/2009/04/20/idUS103880+20-Apr-2009+PRN20090420. ****BAD URL

71. Habitat Heroes website: http://habitatheroes.com/index.html.

72. Fuseworks Media, "Global Warming Fears Seen In Obsessive Compulsive Disorder Patients," Voxy.co.nz (May 6, 2010), http://www.voxy.co.nz/national/psychiatry-congress/5/47523.

73. 1010 Global Website: http://www.1010global.org/.

74. "The infamous No Pressure ad," http://www.mrctv.org/public/checker. aspx?v=hdkU6U2G4z.

75. Damian Carrington, "There will be blood– watch exclusive of 10:10 campaign's 'No Pressure' film," *The Guardian* (September 30, 2010), http://www. theguardian.com/environment/blog/2010/sep/30/10-10-no-pressure-film.

76. The game used to be at: http://www.abc.net.au/science/planetslayer/ greenhouse_calc.htm.

77. Andrew Bolt, "Column—Green slayer is after your children," *Herald Sun* (June 13, 2008), http://blogs.news.com.au/heraldsun/andrewbolt/index.php/ heraldsun/comments/column_green_slayer_is_after_your_children/.

78. Wesley J. Smith, "Radical Environmentalism: Moving Us Toward a Eugenic Culture of Death," *Wellsphere* (February 3, 2009), http://www.wellsphere.com/ bioethics-article/radical-environmentalism-moving-us-toward-a-eugenic-culture-of-death/598322.

79. LiveScience Staff, "Save the Planet: Have Fewer Kids," *Live Science* (August 3, 2009), http://www.livescience.com/9701-save-planet-kids.html.

80. Bryan Nelson, "Was Genghis Khan history's greenest conqueror?" *Mother Nature Network* (January 24, 2011), http://www.mnn.com/earth-matters/climate-weather/stories/was-genghis-khan-historys-greenest-conqueror.

81. David Owen, "Economy vs. Environment," *The New Yorker* (March 30, 2009), http://www.newyorker.com/talk/comment/2009/03/30/090330taco_talk_owen.

82. Ronald Bailey, "Delusional in Durban," Reason.com (December 5, 2011), http://reason.com/archives/2011/12/05/delusional-in-durban.

83. Kelvin Kemm, "Renewables not the solution for Africa," *Engineering News* (December 2, 2011), http://www.engineeringnews.co.za/article/renewables-not-the-solution-for-africa-2011-12-02.

84. Ban Ki-moon, "We Can Do It," *New York Times* (October 25, 2009), http:// www.nytimes.com/2009/10/26/opinion/26iht-edban.html.

85. Li Xing, "Population control called key to deal," *China Daily* (December 10, 2009), http://www.chinadaily.com.cn/china/2009-12/10/content_9151129.htm.

86. Diane Francis, "The real inconvenient truth," *Financial Post* (December 8, 2009), http://www.financialpost.com/story.html?id=2314438.

87. Thomas L. Friedman, "Our One-Party Democracy," *New York Times* (September 8, 2009), http://www.nytimes.com/2009/09/09/ opinion/09friedman.html.

88. Ed Pilkington, "Put oil firm chiefs on trial, says leading climate change scientist," *The Guardian* (June 22, 2008), http://www.theguardian.com/environment/2008/jun/23/fossilfuels.climatechange.

89. John Vidal, "Kingsnorth trial: Coal protesters cleared of criminal damage to chimney," *The Guardian* (Sept. 10, 2008), http://www.theguardian.com/environment/2008/sep/10/activists.carbonemissions.

90. Donald A. Brown, "A New Crime Against Humanity? The Fossil Fuel Industry's Disinformation Campaign on Climate Change," *Social Ecology Institute of British Columbia* (October 2010), http://bcise.com/CurrentIssuePapers/Oct-2010/a-new-crime-against-humanity.pdf. \*\*\* BAD URL

91. Paul Thornton, "On Letters from Climate-Change Deniers," *Los Angeles Times* (Oct. 8, 2013), http://www.latimes.com/opinion/opinion-la/la-ol-climate-change-letters-20131008,0,871615.story#axzz2nyJuucW9.

92. Nathan Allen, "Reddit's science forum banned climate deniers. Why don't all newspapers do the same?" *Grist* (Dec. 16, 2013), http://grist.org/climate-energy/reddits-science-forum-banned-climate-deniers-why-dont-all-newspapers-do-the-same/.

93. Ian Johnston, "'Gaia' scientist James Lovelock: I was 'alarmist' about climate change," *NBC News.com* (April 23, 2012), http://worldnews.nbcnews.com/_news/2012/04/23/11144098-gaia-scientist-james-lovelock-i-was-alarmist-about-climate-change.

94. Mark Halle, "The UNEP That We Want," *International Institute for Sustainable Development*, September 17, 2007, http://www.foxnews.com/projects/pdf/113009_IISDreport.pdf.

95. Wesley J. Smith, *Culture of Death: The Assault on Medical Ethics in America* (Jackson: Encounter Books, 2002).

96. Quote from Ingrid Newkirk: http://www.brainyquote.com/quotes/quotes/i/ingridnewk104090.html.

97. Wesley J. Smith, "PETA's Non-Apology Apology," *National Review Online* (May 6, 2005), http://old.nationalreview.com/smithw/smith200505060923.asp. \*\*\* BAD URL

98. Wesley J. Smith, "PETA's Non-Apology Apology for Odious Holocaust on Your Plate Campaign," *National Review* (May 5, 2005), http://nationalreview.com/human-exceptionalism/326903/petas-non-apology-apology-odious-holocaust-your-plate-campaign.

99. "Van Jones," *Wikipedia*, http://en.wikipedia.org/wiki/Van_Jones.

100. "Universal Declaration of Rights of Mother Earth," from the *World People's Conference on Climate Change and the Rights of Mother Earth* (April 22, 2010), http://therightsofnature.org/universal-declaration/.

101. Visit the *Community Environmental Legal Defense Fund* website here: http://www.celdf.org/.

102. "Pachamama," *Wikipedia*, http://en.wikipedia.org/wiki/Pachamama.

103. Shannon Biggs, "Legalizing Sustainability? Santa Monica Recognizes Rights of Nature," *Global Exchange* (April 11, 2013), http://www.globalexchange.org/blogs/peopletopeople/2013/04/11/legalizing-sustainability-santa-monica-recognizes-rights-of-nature/.

104. "Hydraulic fracturing," *Wikipedia*, http://en.wikipedia.org/wiki/Hydraulic_fracturing.

105. "Ecology in Ecuador," *Los Angeles Times* (September 2, 2008), http://articles.latimes.com/2008/sep/02/opinion/ed-nature2.

106. UNFCCC Ad hoc Working Group on Long-term Cooperative Action under the Convention, "Update of the amalgamation of draft texts in preparation of a comprehensive and balanced outcome to be presented to the Conference of the Parties for adoption at its seventeenth session. Note by the Chair" (July 12, 2011), http://unfccc.int/resource/docs/2011/awglca14/eng/crp38.pdf, pp. 44-45.

107. "Origin Story," The Pachamama Alliance, http://www.pachamama.org/about/origin-story.

108. "Rights of Nature: FAQs," http://celdf.org/rights-of-nature-frequently-asked-questions.

109. See Wesley Smith, "The Silent Scream of the Asparagus," *The Weekly Standard* (May 12, 2008), available online at http://www.weeklystandard.com/Content/Public/Articles/000/000/015/065njdoe.asp.

110. "The dignity of living beings with regard to plants," *Federal Ethics Committee on Non-Human Biotechnology*, http://www.ekah.admin.ch/fileadmin/ekah-dateien/dokumentation/publikationen/e-Broschure-Wurde-Pflanze-2008.pdf.

111. Kate Lindemann, Professor of Philosophy Emerita, Mount Saint Mary College, "Moral Community," http://faculty.msmc.edu/lindeman/mc.html.

112. "Sentient," Merriam-Webster.com, http://www.merriam-webster.com/dictionary/sentient.

113. Guatam Naik, "Switzerland's Green Power Revolution: Ethicists Ponder Plants' Rights," *The Wall Street Journal* (October 10, 2008), http://online.wsj.com/article/SB122359549477921201.html.

114. Michael Marder, "If Peas Can Talk, Should We Eat Them?" *New York Times* (April 28, 2012), http://opinionator.blogs.nytimes.com/2012/04/28/if-peas-can-talk-should-we-eat-them/.

115. Visit Michael Marder's personal website here: http://www.michaelmarder.org/.

116. Linton Weeks, "Recognizing The Right of Plants To Evolve," *NPR* (October 26, 2012), http://www.npr.org/2012/10/26/160940869/recognizing-the-right-of-plants-to-evolve.

117. Natalie Angier, "Sorry, Vegans: Brussels Sprouts Like to Live, Too," *New York Times* (December 21, 2009), http://www.nytimes.com/2009/12/22/science/22angi.html.

118. "Draft Ecocide Act," *Eradicating Ecocide*, http://eradicatingecocide.com/overview/ecocide-act/.

119. No longer available at the original website, but it can be viewed at http://web.archive.org/web/20100818035651/http://www.thisisecocide.com/thesolution/.

120. "Ecocide—A Crime Against Peace," http://www.youtube.com/watch?v=-47A8yWghbs.

121. "Draft Ecocide Act," *Eradicating Ecocode*, http://eradicatingecocide.com/overview/ecocide-act/

122. Ibid.

123. Devin Rawlinson, "Mock trial finds Tar Sands spill 'bosses' guilty of ecocide," *The Independent* (October 1, 2011), http://www.independent.co.uk/environment/green-living/mock-trial-finds-tar-sands-spill-bosses-guilty-of-ecocide-2363988.html.

124. "After the Trial, the Sentence," *The Hamilton Group*, http://www.thehamiltongroup.org.uk/common/ecocide-sentence.asp.

125. "Justice Restored: Bannerman and Trench Sentence for Ecocide," Youth-Leader.org *Magazine*, http://www.global1.youth-leader.org/2012/04/justice-restored-bannerman-and-trench-sentenced-for-ecocide-3/.

126. "Ecocide: a good example!" *European Platform Against Windfarms*, http://www.epaw.org/multimedia.php?lang=en&article=g22.

127. Marcin Gerwin, "Healthy Planet and the Law of Ecocide – an Interview with Polly Higgins," *The Permaculture Research Institute* (April 18, 2013), http://permaculturenews.org/2013/04/18/healthy-planet-and-the-law-of-ecocide-an-interview-with-polly-higgins/.

128. Juliette Jowit, "British campaigner urges UN to accept 'ecocide' as international crime," *The Guardian* (April 9, 2010), http://www.theguardian.com/environment/2010/apr/09/ecocide-crime-genocide-un-environmental-damage.

129. "Remediation," BusinessDictionary.com, http://www.businessdictionary.com/definition/remediation.html.

130. "Reclaiming Alberta's Oil Sands," *Government of Alberta*, http://environment.alberta.ca/02012.html.

131. David Sims, "Will We Ever See A Technology for Cleaning Up Oil Sands Tailing Ponds?" *Thomasnet News* (April 11, 2013), http://news.thomasnet.com/green_clean/2013/04/11/will-we-ever-see-a-technology-for-cleaning-up-oil-sands-tailing-ponds/. *** BAD URL

132. "Endorsements," *Eradicating Ecocide*, http://eradicatingecocide.com/supporters/endorsements.

133. "Existing Ecocide Laws," *Eradicating Ecocide*, http://eradicatingecocide.com/overview/existing-ecocide-laws/.

134. TreeHugger, "New Zealand Grants a River the Rights of Personhood," *Care 2 Make a Difference* (September 8, 2012), http://www.care2.com/causes/new-zealand-grants-a-river-the-rights-of-personhood.html.

135. Carl Cohen, *The Animal Rights Debate* (Roman and Littlefield, 2001), 17, available online at http://carl-cohen.org/books/AnimalRightsDebate/chapter3.pdf.

136. David S. Oderberg, "The Illusion of Animal Rights," *Human Life Review* (Spring-Summer 2000), 42, available online at http://www.scribd.com/doc/50084724/Oderberg-The-Illusion-of-Animal-Rights.

137. Louise Gray, "David Attenborough—Humans are a plague on Earth," *The Telegraph* (January 22, 2013), http://www.telegraph.co.uk/earth/earthnews/9815862/Humans-are-plague-on-Earth-Attenborough.html.

138. Population Matters, "Is humanity a plague?" *Population Matters* (January 24, 2013), http://webcache.googleusercontent.com/search?q=cache:http://www.populationmatters.org/2013/population-matters-news/humanity-plague/.

139. "The most surprising demographic crisis," *The Economist* (May 5 2011), http://www.economist.com/node/18651512.

140. Edwin Black, *War Against the Weak* (Four Walls Eight Windows, 2003); also see http://www.waragainsttheweak.com/.

# INDEX